BRAIN
LINES
™

Don't worry about worry-lines –
they're not wrinkles, they're Thinkles!

FOREWORD

The origin of all naturalistic phenomenon in the human world relates to the effects of a complex interplay between molecular (neuronal) biology and environmental influence. Neuropsychological processes or tendencies are hardwired into each person, present from birth and modified throughout life to some degree. The outward display of any human behavior marks the interaction of the DNA-influenced neuronal interactions with the ongoing environmental stressor(s) of that time. The book, **BrainLines***, adds a physical "twist" to this naturalistic principle just as handedness previously was scientifically predicted and shown to delineate brain functional dominance or dominant tendencies.*

Using the concept of BrainLines, there is much to learn about our society and its humanistic groups via further scientific studies on grouped subjects who share the same and different BrainLine markings. Neuroscientists, Neuropsychologists, Neurophysiologists and animal behavioralists should easily be able to adapt BrainLine signs to the experimental method using brain imaging and quantifiable testing techniques.

It should come as no surprise that another physical marking has physiological and neurobehavioral relevance. The skin-central nervous system organ (i.e. what developmental neuroscientists refer to as the neuro-ectodermal layer) is of similar origin in embryonic human development. BrainLine markings give additional relevance to this well-known neuroembryological concept. The description and relevance of these unique forehead markings, the BrainLines, should evoke a strong sense of intrigue in those scientists who study human personality, behavior and psychosocial dynamics.

Jonathan Artz, M.D., Clinical Neurologist and Neurophysiologist

TABLE OF CONTENTS

Preface i

Introduction v

Quick Profile Summaries 1

Chapter One:
 BrainLines: Genetic Markings and Inner Knowledge 7

Chapter Two:
 Split-Brain Research 23

Chapter Three:
 Applying the Knowledge 37

BrainLines Personality Profiles:
 Balanced Double 49
 Center 61
 Clear 75
 Left Clear 89
 Left Double 103
 Right Clear 117
 Right Double 131

Bibliography 145

About the Authors 149

PREFACE

When I was 37, I began to notice my wrinkles. My face was a physical record of a life spent in the California sun. Disturbing, yes, but what could I do to make myself feel better about them?

My solution was to draw a self-portrait accentuating every wrinkle on my face. While drawing my forehead, I became curious as to why the 'frown lines' between my eyebrows were of different lengths. I saw a deep, long vertical line on one side and another deep vertical line, only half as long, on the other. Why were they different from each other? Why were they so much deeper than my other wrinkles? Were they caused by always sleeping on one side? No.

From that moment on, I was fascinated by these wrinkles. Enthusiastically, I plunged into research and found that there were multiple formations. Some people, even the elderly, had no pattern of lines. Some children, even toddlers, had deep lines. Today I call them *BrainLines*.

Comparing my friends' BrainLines with what I knew of them, I began to see relationships between the differences in their lives and their BrainLines patterns. I began to find similar personality traits in people with similar 'readings.'

During the 1960's and '70's, the scientific research into left and right brain hemisphere functions and how they influenced behavior became popularized. I began to wonder if there was a correlation between BrainLines and the personality. When I started thinking in terms of left-brain and right-brain processes, as defined by researchers, they became an effective way to describe the behavior patterns of my friends and myself. I had stumbled across the BrainLines connection! I noticed that my 'right-brained' friends had a pronounced vertical frown line on the left-hand side, while my more 'left-brained' friends had one on the opposite side.

So, my life began to change as I spent more and more time interviewing people. I would read people's BrainLines and ask them pertinent questions about their lives and their perceptions. Soon, I could tell them their basic personality traits. The accuracy of these basic readings made people think that I was psychic. I felt I was really on to something, a method of reading everyone I met, even strangers on the street.

Eventually, it became apparent that there were seven basic personality types that were discernible from BrainLines patterns. By combining this knowledge with the solid scientific research into left- and right-brained behavior traits, I realized I had found a personality identification tool.

BrainLines has developed into an extremely efficient, emotionally satisfying method of helping the average person relate to their inner and outer worlds. Men and women were so surprised how accurately I could read them, and I am so gratified when young women smile and thank me for helping them feel better about themselves.

Through the years of research and observation, I was encouraged by the example of another pioneer in the typology field, Katharine Briggs. Without formal training in psychology, she developed a system which later became the respected and widely-used Myers-Briggs Type Indicator. Her courage and ability to fill a human need were an inspiration to me. When I asked Dr. Randy Thornhill, Distinguished Professor of Biology at the University of New Mexico, if my research findings and profile system would be taken seriously, he told me that the development of a system in isolation, free from the academic world, gives it strong accreditation.

Some skeptics have said they find any kind of personality typing restrictive, that I'm trying to fit people into little boxes. A BrainLines reading is not a box, but a launching pad for the preferences and talents with which we are all born. As the great

Swiss psychiatrist Carl Jung said, human behavior is not random, but predictable and therefore classifiable.

By looking at a person's face I feel universally connected with a stranger. With my better sense of knowing my own strengths and weaknesses as a Left Double, I have greater insight into the old questions, "Who am I?" and "Who are you?"

Discovering BrainLines has been a momentous addition to my life. But, were it not for my partner and writer, Paul Reffell (also a Left Double), BrainLines would not be recorded. He has played a major role in the development of the BrainLines system. It gives us great pleasure to now share this creative collaboration with you and may it give you the comfort and insight it has given both of us.

Donna Oehm Sheehan, Evolutionary Behaviorist

INTRODUCTION

The fear of aging and the commodification of their bodies make women prone to the hard-sell for cosmetic **Botox** injections. Women spend two billion dollars a year to have their faces paralyzed to take away the lines between their eyebrows. It's a temporary, expensive and possibly life-threatening ritual, since botulinum toxin is one of the most toxic substances known.

What if just a portion of that money were available every year to fund women's health, education and empowerment programs around the world? What if that money were used to prevent violence against women? What if, instead of getting a temporary 'fix' for worry lines, we learned to recognize their true worth? What if they weren't wrinkles, but **"thinkles"** that reveal the balance of the left and right sides of our brains?

Donna has spent forty years all over the world doing independent research on the correlation between these forehead lines and personality. She asked everyone her "twenty questions" and discovered seven basic formations that relate to individuals' left-and right-brain traits. We call them **BrainLines**. They are the result of the neuromotor influence of the two sides of the brain on the musculature in the center of the forehead.

The seven basic **BrainLines** formations are guides to seven personality profiles that reveal each person's traits; their left-brain and right-brain propensities. And they cross ethnicity, race, nationality and gender lines.

So now, instead of worrying about worry lines, we can 'read' people in everyday life, in relationships, business, politics, law and medicine, to understand and relate to everyone around us. We can tell what our kids' greatest talents might be, why someone works in a completely different way from us, why we are so compatible with one person and not the next, how someone's perception of the world will differ from ours, what work would be most rewarding for us.

BrainLines are physical markers of personality, easily seen and easy to read. A **BrainLines** reading is a form of typology that doesn't require learning about astrology, numerology, enneagrams, palmistry, Jungian archetypes or the Myers-Briggs system. It is intuitive, because **Brainlines** are physical markers that have always been right there in front of us, species-wide, to inform us about each other. We simply forgot their importance and succumbed to the male-driven cultural bias against them, considering them 'ugly' signs of aging.

Let's stop feeding the system that makes women feel bad about themselves, learn to appreciate and use **BrainLines**, and work on empowering girls and women for the future.

There are no bad **BrainLines**!

QUICK PROFILE SUMMARIES

Use these quick summaries to get a feel for what BrainLines can tell you about yourself and the people you know. For all the information you need to know about the brain function, temperament, social ability, relationship needs, compatibilities and working strengths of each reading – and to find out which famous people share your BrainLines pattern – go to the Full BrainLines Profiles, which start on Page 49.

BALANCED DOUBLE – *(full profile on page 49)*

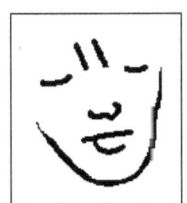 Scattered? Sure. Frenetic? Maybe. Multi-talented? You bet! You have an idea-a-minute, from the abstract poetic to the hard edge of sequential logic. All options are open to you, which makes you a great problem-solver if you can focus long enough. Some of us can't see all sides as you do. As you go out in the field, sniffing every scent, finding out all the information, others will be following the footpath. Both may reach the same point, but you will have discovered so much more along the way.

Relationship tip: You're juiced by happiness, and will bend over backwards to stay that way. Such flexibility is like money in the bank when it comes to a relationship. Yet you don't wimp out when times are tough. You find every possible solution. When you choose the perfect partner, you will be an inspiration to them with your whirlwind of ideas and interests. Two *Balanced Doubles* together? Anything is possible!

CENTER – *(full profile on page 61)*

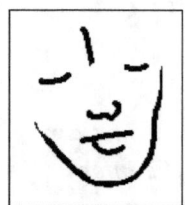

Did anyone ever tell you they thought you were reading their mind? Have you ever felt like you knew the answers to questions before they were asked? No surprise. You are equipped with a DSL connection to your intuition, which means your hunches sometimes surprise even you. You've also got a knack for finding a logical way around life's challenges. You can do it all, if you can let go of the reins on your mind and trust your instincts.

Relationship tip: You're the most interesting person you know, and you can take love or leave it. But once you find the person who can keep you stimulated on all levels, you can dig deep into your reserves of love and compassion. You'll need your alone time, but don't overdo it. Your partner will probably want more shows of affection than you anticipate. Give it a whirl, you'll be making both your lives all the richer.

CLEAR – *(full profile on page 75)*

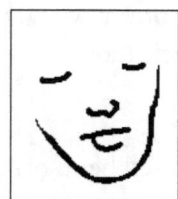

You'll never feel the need for cosmetic surgery between your eyebrows. Maybe that's how you can be so focused on your goals. On the way, you're appreciating all the trappings of success for which you've worked so hard. You're designed to adapt to any circumstances, but, in adulthood, beware of becoming a hard-liner. You've left some egos floating in your wake, but it's a tough world out there and you'll think about the meaning of life once you've got where you're going.

Relationship tip: You're looking for the ideal mate in a stable relationship. The less emotional, the better, as far as you're concerned. Growing up, some dates may have called you insensitive, so you've learned to shape your responses to anyone's needs. The pressure of doing so will build. Better tell your partner how you feel before you explode. With your extensive abilities, you're a prize catch for any prospective partner.

LEFT CLEAR – *(full profile on page 89)*

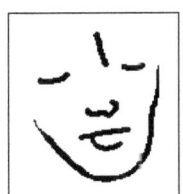

You're a poet with a surgeon's eye. You are most successful in imaginative, even idealistic, pursuits. If you're an artist, you're the rare kind that has the confidence and pragmatism to market their own work. If you're an entrepreneur, you want to make people's lives better with your inventions, even while making a living from them. Don't hide your soft side, there's room at the top for a big heart.

Relationship tip: "I'm sorry, it just feels like we're getting in a rut." Did you say that? Love is your bed of oysters and you want to find the pearls. You've been disappointed when your vision of a relationship didn't come true. Once you've found someone who can keep up with your need for excitement, you'll devote yourself to refining the relationship until it's running smoothly. Then you'll be able to work towards your dream. Half the fun is getting there.

LEFT DOUBLE – *(full profile on page 103)*

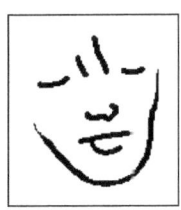

"There are no pastimes, only passions." Bet you wish you'd said that. If you've been given a free enough rein in your lifetime, you feel the truth of that saying. There's a neon sign over your head that reads, "Creative." And that can cover everything from painting through writing to social skills and lifestyle. You are aware of all the emotional levels around you. If you could, you'd ask everyone the questions that would lay their souls bare. You are spontaneous, sensitive and emotional. You're quite a handful!

Relationship tip: Courtship, to you, means mystery and theater. Not to mention passion, which is a constant in your life. Your partner will have to be imaginative to keep up with you. You also need someone who is not intimidated by non-conformity and creativity, and who is not easily embarrassed. You may have to "upgrade" partners a few times, but the search will be exciting, delicious and full of fun. Just like life itself.

RIGHT CLEAR – *(full profile on page 117)*

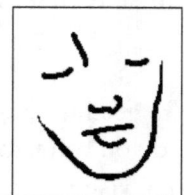 The speed of your mind sometimes leaves others in the dust. You make quick decisions based on facts, and you don't let emotion slow you down. That cool exterior doesn't fool your closest friends and family, who know the love and compassion of which you are capable. So what if you've never been the life of the party, the world needs your laser-like ability to cut through the hype. You are a lighthouse in a stormy sea: dependable, safe, solid and true.

Relationship tip: Remember, it's O.K. (but not mandatory) to lighten up with the people who love you. You can show love in many ways other than programming the VCR or making picture-perfect pies. The right relationship must be as important as your career. Make allowances for the unexpected while drawing up the plans for your life together.

RIGHT DOUBLE – *(full profile on page 131)*

 You're the observer with the twinkling eye and the concise one-liner to sum up the occasion. You could be a rocket scientist with an art degree. Although you would love to organize your life, you have that quirky side that keeps everything slightly off-balance. Don't deny your sensitivity to the feelings of the people around you. It's a valuable asset. Thank goodness you're around to give sensible advice to the more illogical among us, yet maintain your sense of humor. If you were an IPO, you'd be a great investment.

Relationship tip: You are tolerant, idealistic, rational and practical. Need we say more? Well, O.K., you're also prone to letting your career take precedence over your relationship. But you have some built-in emotional radar that will tell you when your partner's getting antsy for expressions of love. Pay attention and don't keep your love-life on auto-pilot.

Before Rob and Jenny met, each of them was focused on a spiritual path, asking themselves, "Who am I? What is the meaning of life?" After many years of a rich but unsettled existence, they finally found each other. Rob is eighteen years younger, but their similar BrainLines and the merging of their psyches made it easy to commit to many long years together. They have a loving, conflict-free collaboration, even though they have separate bedrooms.

CHAPTER ONE

BRAINLINES: GENETIC MARKINGS & INNER KNOWLEDGE

Troth, sweet lord, thou hast a fine forehead.

Shakespeare: *Troilus and Cressida*

Take a look at yourself in the mirror. Now frown slightly. If you see vertical lines appear between your eyebrows, you are looking at your *BrainLines*. You may have been acutely aware of them before, or you may have ignored them, but now we want you to see those lines in a whole new light. Most people think that they are simply the lines that form over years of squinting or frowning and so have dismissed them as a sign of aging.

Closer study will reveal, however, that the patterns differ from person to person, even though we share similar musculature in that area. You've probably had the same lines since birth. These lines are our personal indicators of brain hemisphere dominance in our thought processes and personality.

Historically, they have been called frown lines, worry lines, concern lines or just plain wrinkles. Now you'll know them as BrainLines. They are the indicators of the way your brain was genetically programmed to function. That programming determined whether the left and right sides of your brain work in balance, or whether one is dominant over the other.

The traits enabled by your unique mix of brain functions have laid the groundwork for the rest of your life. Everything you have learned, all the sensory input you have received, every experience you have had has built upon that genetic foundation. Yet, beneath all your learned behavior, the foundation remains, governing your preferences, emotions and reactions to everything and everyone you encounter.

BrainLines are created by the motor functions of the nervous system, controlled by the brain. The fact that BrainLines are displayed on opposite sides of the center-line of the brow suggests that they are created by the separate hemispheres, with their asymmetry showing the preponderant influence of one side or other of the brain.

If there is a pronounced dominance of one hemisphere or the other, it shows up in the relative length and depth of the BrainLines. This explains why someone with a left hemisphere-dominant personality has a longer line on the right-hand side (the side controlled by the left hemisphere) of the *glabella*. The glabella is the anatomical name of the bone underlying the area between the eyebrows. It will sometimes be used in the book as a synonym for the area in which the BrainLines are found.

No, The Other Left

The terminology of left and right can be confusing when applied to physical characteristics. We call the eye on our left-hand side our left eye. BrainLines are classified according to the side of the individual's face on which they appear. Each hemisphere of the brain governs the opposite side of the body, so the left brain controls the right-hand side of the body and the right brain controls the left-hand side.

So. BrainLines are named after the side of the brow on which they appear, but are markers of the influence of the opposite side of the brain. Thus, the Left BrainLine is located on the left-hand side of the glabella, the side controlled by the right brain. Similarly, the Right BrainLine is located on the right-hand side, controlled by the left brain.

For now, take a look at the line-up of basic patterns. See if you can guess which hemisphere influences the lines in the drawings. Here's a clue - the face in the middle shows left-brain dominance.

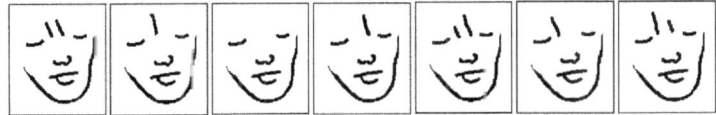

Don't worry if you haven't got the hang of it yet. These faces will be further explained later in the chapter.

A New Look at an Old Wrinkle

Until now, BrainLines have been seen by the more cosmetically-conscious among us as merely troubling signs of age, as frown lines, concern lines, wrinkles, and furrows.

But BrainLines are more than just the result of late-night squinting at reports due next day or of spending beach-time with eyes narrowed against the sun. Someone may have been a lifetime surfer, squinting in the sun and spray, yet display no distinct lines between the eyebrows. That surfer would be what we call a Clear, which we count as one of the seven readings.

For thee rest of us, our BrainLines are with us from birth as genetic markings. If you look at photos taken in your childhood in which you are squinting or crying, you will see that these lines have been with you all along. They were not as deeply etched but were, nevertheless, a part of your physical structure. Now they no longer need be regarded as a problem for a cosmetic surgeon to fix with implants, Botox injections or a scalpel. BrainLines have not suddenly appeared on your face since you started worrying about your age!

At this moment, there are thousands of women, and some men, who are seriously considering having their BrainLines removed. We hope they will think twice before going through the mental and physical suffering involved in surgically hiding their personalities. Not to mention the possible side effects. We quote from a Botox ad:

"Problems breathing, swallowing or speaking. These problems happen hors, days, to weeks after an injection...Death can happen as a complication.....Swallowing problems may last for months...may need a feeding tube...Spread of toxin effects...may affect areas of the body away from the injection and cause symptoms of botulism....muscle weakness all over the body, double vision, drooping eyelids, hoarseness, trouble saying words clearly, loss of bladder control......dry mouth, tiredness, headache, neck pain, eye problems, urinary tract infections, allergic reactions....."

Yikes!!

When we met Julie and told her about BrainLines, she said, "Oh my God! I had them removed! I had collagen injections. If I'd known, I could have saved myself a lot of money." She told us she was a Center. She remembered the single line that she had always tried to cover with long bangs because the kids at school used to call her "Buttface."

BrainLines are genetic markings which are visible in newborns, and are fixed from about the age of two years, when the skull bones are less mobile and the forehead and nose have developed their unique forms. In order to see a BrainLines reading on young children it may be necessary to very gently press the eyebrows towards the glabella. Too much force will push the skin into unnatural folds, but gentle pressure will show the BrainLines pattern that will become permanently visible later. Until the mid-twenties, they may only be apparent when frowning, becoming more pronounced over the succeeding years as the skin loses its elasticity.

Hope Springs Internal...

It has become ever more apparent from genetic research that all humans are fundamentally alike. Despite the various skin-tones and fatty deposits that make us look different and which some believe are "just cause" for prejudice, we all share the basic attributes that make us human. One of those attributes is our BrainLines pattern, which is seen in all races and cultures around the world. Everyone has the markers of their powerful and versatile brain, the clearing-house for input from the outside world and also the organ that creates our thoughts and emotions.

Physically, the brain is a double organ, with the left and right halves (or hemispheres) being connected by neural pathways. *There is no central information-processing center; there are two brains working side-by-side.*

In the mainly right-handed Western culture the left half of the brain is usually considered the dominant hemisphere because it controls the verbal skills (both written and oral) and is associated with the more linear and analytical modes of processing information. The right side of the brain imparts the more visual, intuitive and integrative thought processes. The way in which the two hemispheres of the brain deal with sensory input affects cognitive processes and behavior patterns.

The personality is the combination of qualities from each hemisphere of the brain and is unique to each individual.

A Puzzle Solved

When Donna Sheehan first noticed that there were pronounced differences in people's "frown lines" it was simply an anatomical puzzle, until these differences were found to relate to people's psychological profiles.

It became apparent that when someone, for example, fitted the verbal, analytical, methodical, linear thinking style of the left-brain-dominant personality, the BrainLine on their right-hand side was longer than the other line, or was the only line. The same was true of the BrainLine on the left-hand side in the case of the right-brain-dominant person. Once this link was established, a startlingly accurate correlation of the lines with the descriptions of left- and right-brain traits became apparent.

Patterns of Personality

There are seven basic BrainLines readings. Most people display one of these patterns, although there are variations and multiples. You will soon have a grasp of the basic seven and the profiles that accompany them. Take another look at the faces and match them with their names.

Here are the basic seven:

BALANCED DOUBLE: Two lines of equal length, one on each side of the area between the eyebrows. Balanced

Doubles combine the qualities of the left and right brains in equal measure.

CENTER: A single vertical line in the center of the area between the eyebrows. The line will sometimes be to one side of the center, but the lower end of the line does not turn to the left or the right. Balanced Centers can access both sides of the brain at once or either side, at will.

CLEAR: No clearly-defined pattern of vertical lines on the area between the eyebrows. They may show a jumble of partial lines, dimples and swirls, or no markings at all. The Clear person feels influence from both sides of the brain.

LEFT CLEAR: Single line on the left-hand side of the area between the eyebrows. The Left Clear person combines right brain and Clear influences, with the right brain being stronger.

LEFT DOUBLE: Two lines, one on each side of the area between the eyebrows. The line on the left-hand side is longer than the one on the right-hand side. The Left Double person combines the left and right brain influences with the right being the stronger.

RIGHT CLEAR: Single line on the right-hand side of the area between the eyebrows. The Right Clear person combines left brain and Clear influences with the left brain being the stronger.

RIGHT DOUBLE: Two lines, one on each side of the area between the eyebrows. The line on the right-hand side is longer than the one on the left-hand side. The Right Double person combines the left and right brain influences with the left brain being the stronger.

The conceptual platform of the BrainLines profiles is linked to research into the relationship of the brain hemispheres to behavior by Robert Ornstein, Michael Gazzaniga and others. The BrainLines system defines seven distinct ways of receiving, processing and expressing information.

During the past fifty years, we have collected survey data from a wide variety of people, representing diverse ethnic and economic backgrounds and both sexes.

The population percentages of each BrainLines reading are roughly as follows:

Balanced Double	23%
Center	19%
Clear	32%
Left Clear	6%
Left Double	7%
Right Clear	5%
Right Double	8%

The numbers give an idea of the possible influence of the first three readings in our society. The Clear reading is an especially powerful force in all levels. We become so used to working with members of the top three readings that, when we come across the other personalities, we can be at a loss as to how to work with them. That is where the knowledge of inherited preferences can be so useful.

I Am What I Am...

In genetic research there is conclusive evidence of hereditary influence on many processes, including brain function. From the moment of conception, we each have a mixture of cerebral traits which form the foundation for our individual personalities. Although we share the same basic brain structure, our brains are each programmed differently.

Thus, the fundamental differences between us have more to do with the genetic arrangement of our brain-cell activity than with culture, race and gender.

Desirable Signs

When a trait affects the genetic profile of a species, there is usually a physical marking which displays that trait. A physical marking supports the continuation of a species by allowing the opposite sex, at a glance, to pick the mate with the most beneficial qualities. The male frigate bird with the biggest, reddest throat pouch is the most desirable as a mate, not because of the pouch itself, but because it is a sign of the bird's health and genetic strength, which is of value to the species. The female frigate bird instinctively recognizes this genetic vigor and is attracted to it.

The human's world is shaped by our intellect. Our brains have given us the power to survive under almost any conditions by inventing ways to create shelter, feed ourselves and produce things and ideas to keep our brains efficient as tools for the survival of the species. So many mental skills have been needed for our continued survival that it seems only natural that there should be some physical marking that displays the way our brains work.

BrainLines, as genetic indicators, are as ancient as the human race itself. Our earliest frescoes, paintings and sculptures show them clearly. They are part of the face that we take for granted, but they may play an instinctive and unrecognized role in the way we choose our friends and partners. BrainLines have always been there for us to see, but not until now do we finally understand their significance and the natural logic of their existence. As the old saying goes, "The obvious is the last thing we see."

Sylvia, a Clear, met George at the office Christmas party. They worked in different sections of the corporation and had never seen each other before. Sylvia noticed George looking at her over the rim of his complimentary glass of Zinfandel and liked the way he looked. Before starting a conversation with George, she read his BrainLines and decided that, even though she sometimes had difficulty in relating to Left Clears, the party was getting dull and she needed a little challenging conversation.

BrainLines, the natural physical markings of personality, are a common bond, a signal of humanness, regardless of race, sex or nationality, creating a greater understanding between us. We all want approval and understanding. It is empowering to be able to

16

walk anywhere in the world among other people and feel the sense of deep communication BrainLines can give.

You Just Have To Know Where To Look

As it became obvious that BrainLines really are the genetic markings of the cerebral functions and their effects on the personality, the question arose, "How do they form?"

BrainLines are the result of the muscle actions of the glabella area, controlled by the motor functions of the left and right sides of the brain. We believe that if one side of the brain is dominant, that side creates a longer, deeper line on the side of the glabella opposite to it when we frown. To repeat, the line is on the opposite side because each side of the brain governs the opposite side of the body.

We each have different innate psychological propensities which are of paramount importance in interpersonal relations. The most logical place to physically display them would be a highly visible place, uncovered by hair growth. The area that fits this description most fully is the forehead, near the eyes, which are the focal point of initial contact between humans.

What's the Use of BrainLines?

This book will show you how to recognize and read BrainLines quickly and easily, giving you the basic, underlying psychological tendencies of the person you are reading, including yourself. BrainLines will not tell us how to read people's minds, but will show the fundamental cognitive preferences in ourselves and others.

BrainLines will give you insight into someone's thought processes, what kinds of things might interest them, how they perceive the world, to whom they are attracted and what their reactions to certain situations might be. This is enormously useful in every personal encounter, whether social or professional.

The ability to understand ourselves that comes from the knowledge of BrainLines is a very empowering addition to our lives. When you can look at yourself in the mirror, see what psychological propensities you were born with and compare them to how you have lived your life, you may find yourself realizing which life-altering decisions were your own and which were imposed upon you. You can find ways to change yourself by using your strengths and forgiving yourself for your weaknesses. You can see how the differences between people make us all stronger, like bricks in an arch. You can pick out the people in a crowd with whom you identify. You can discover the mysterious workings of minds that see the world differently from yours. You can begin to eliminate blame from your relationships with yourself and others.

The Desire For Inner Knowledge

The impetus for research into the many different methods of character analysis came from a desire for true inner knowledge. All behavioral studies address the need for connecting with each other in a clear, honest and intimate way, enhancing the possibility of each person acting in a manner that is harmonious with their nature.

For centuries, the quest has been conducted for the best method of deciphering the mystery of the psyche. Long before the word *psychometrics* was invented to describe the science of psychological testing there was a craving for the kind of insight into the self that is gained by study of the personality and thought processes.

Through the ages, humans have created divinatory methods and belief systems in the search for self-discovery. Most religions pursue connection of the soul or spirit to the particular force of creation they revere. As part of the preparation for that connection the soul must be examined and refined in order to be worthy. In the process, the inner self becomes clearer to the seeker. Even if connection is never attained, the knowledge gained brings a degree of enlightenment.

The Third Eye, The Sixth Chakra

The discipline of yoga is one such pathway to enlightenment. The area of the face in which BrainLines are found is referred to in yoga meditation practices as the site of the *third eye*, of great importance as a symbol of the seat of individual consciousness and wisdom. Through the third eye the yogi can see his true form in the universe. Yoga practitioners believe that there are seven *chakras*, power centers which are arranged vertically in the body from the genital area to the top of the head, through which spiritual energy can be channeled. The sixth chakra, called *ajna*, is situated between the eyebrows at the site of the third eye (and the BrainLines), and is called by some the center of command. Ajna is considered the site of individual consciousness and the meeting-place with the Divine. A dot painted on the site of the third eye signifies peace and dignity.

It has been the practice for some yoga masters to paint lines similar to BrainLines on the glabella. Is that practice rooted in an unconscious awareness of the importance of BrainLines to someone who wishes to display the signs of his powers of insight, spirituality and leadership?

There have been methods for categorizing personality in every major culture since pre-history. Astrology incorporates the influences of planets, stars and the four basic elements in defining personality. Indian and Oriental traditional medicine utilize personality typology as part of their diagnostic and healing regimes. Fortune-telling practices are common to all cultures and many of their personalized prophecies are based on some knowledge of the person's personality.

In the present day, the knowledge of someone's psychological propensities is a powerful tool for commerce, politics and just plain everyday personal contacts. The BrainLines method is a quick and easy way of gaining this knowledge and, unlike other psychological tests, is available for all to use without the need for lengthy questionnaires or examinations. Anyone can use it at will, whenever they meet somebody new, in any circumstance and without intruding upon the other person.

What's My Line?

BrainLines will not act as a record of psychiatric problems, childhood trauma, criminal record or even whether a person will get nasty after a couple of drinks. The circumstances of an individual's history, the type of family life, the quality of upbringing, education, peer influence and pressures to perform well in a chosen career create learned behaviors. BrainLines do not indicate behavior patterns superimposed on the basic genetic foundation.

However, the knowledge gained from a BrainLines reading may be an important factor in efficient interpersonal communication.

"My husband and I have similar BrainLines readings, thank goodness. There's not a day that goes by that I don't feel really thankful that we found each other. We have some kind of primal linkage, like perceiving the world the same; politics, morality, humor, even movies and literature. We come from different countries, very different family backgrounds and there's a twenty-year age-gap between us. But our collaboration in life is so easy and creative that now we are writing books and screenplays together, without once threatening divorce! Life is good!"

Rex and Tracy L.

BrainLines can help to decipher the causes behind some stress-related syndromes by providing a reading of the basic personality profile. That reading can then be used to decide where the stress could be relieved by utilizing more of your strengths.

Learning For Life

Most people think from time to time that they have not lived up to their potential. You may feel that life has been disappointingly unfulfilling or a constant struggle, a Sisyphean contest against unseen forces that seem bent on keeping you down. You may have allowed yourself to make life-choices that were inappropriate, without consideration of your innate strengths and weaknesses, causing you to perform inadequately. You may unconsciously feel the strain of doing things in a way alien to your nature.

Go and look in the mirror, take a look at your BrainLines, then sit down and read the profiles and learn more about yourself and those around you. With the knowledge of BrainLines you will find that your perceptions of the people you know may change and you will find it easier to interact, feel comfortable with anyone and make new friends.

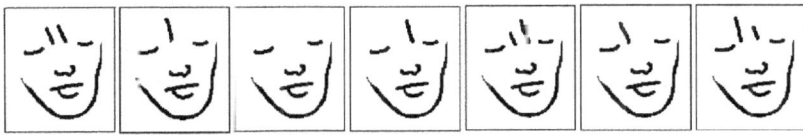

Richard's classes in Constitutional Law were always well attended, attracting many students from other disciplines. As a Center, he had a firm, analytical grasp of the issues and was able to deliver complex information clearly and with a flair that made it entertaining and enlightening. His discussions of ethical concerns went far beyond the parameters of the law as written, opening students' eyes to a wider vision of the moral impact of legal decisions. Richard's classes became the inspiration for many political and legal careers founded on fairness and concern for humanity.

CHAPTER TWO
SPLIT-BRAIN RESEARCH

There is written in your brow, Provost, honesty

Shakespeare: *Measure for Measure*

Behavioral scientists have for a long time studied the dilemma of the two-sided human brain and the fact that each hemisphere of the brain could function alone, with certain mystifying exceptions.

René, Franz et A.L.

There have been some interesting theories on how the brain's two hemispheres are able to process input, and how the brain forms our personalities.

René Descartes, in 1649, theorized that the pineal gland was the unifying organ between the two hemispheres of the brain and the site of the interaction of the soul with the body. He chose it because it is one of the few singular organs in the skull.

Descartes' theory was discredited because, in the opinion of his peers, the pineal gland's construction was too simple to be the site of anything as complicated as the soul's interactions with the senses. What a coincidence that he should have chosen this area as the seat of consciousness, since it corresponds to the "third eye" and the sixth chakra, or power center, of Hindu and Buddhist meditative practices.

Descartes' proponents searched the structure of the brain to find another acceptable singular organ, but could not reach any satisfying conclusions. There was speculation about the corpus callosum. It is the bundle of fibers that connects the hemispheres structurally and allows transmission of sensory input between them, but it was also considered unsuitable. The mystery of how the two cerebral hemispheres were able to take dual inputs and turn them into one concept continued to baffle the medical and philosophical establishments.

Later, there were discoveries and theories that began to consider the left and right brains' functions in the personality. In the early nineteenth century Franz Joseph Gall published his map of the soul. His diagrams showed the surface of the brain divided into small areas, each serving different faculties. The theory was that each separate area could only do one job (such as parental love or conscientiousness) and that each was duplicated in the left and right hemispheres, the brain being completely symmetrical. Gall thought that each hemisphere functioned separately, one taking over from the other when it wore out or was injured.

In 1844, A.L. Wigan, an English physician, put forward his theory of the existence of two separate and complete minds within the human skull. Wigan stated that each hemisphere functioned separately but synchronously, and, if one of them should gain ascendancy over the other, madness would result!

The Split Brain

During the nineteenth century there was extensive research carried out on split-brain patients––those whose cerebral hemispheres had been separated either surgically or through trauma.

These experiments were valuable in discovering the roles played by each hemisphere in thought processes. Through experiments on patients with such cerebral disorders as *alexia without agraphia* (the inability to read or copy written material while retaining the ability to write when given verbal instructions), the functions of various areas of the brain were mapped out. In such cases, lesions would be found in the brain that prevented visual stimuli from reaching the reading center of the left hemisphere. The patient could see the words and write them when they were dictated, but could not read what he had written. Great strides were made in brain research through the study of such unfortunate people.

Researchers found that the left brain hemisphere normally appeared dominant over the right hemisphere. Success in Western culture was measured in business acumen, scientific achievement, literary skills and political ability. All these are predominantly left-brain-based skills, and the European school system was geared around them. The area governing spoken-verbal abilities was found to be normally situated in the left brain. Also, since most people were right-handed, the left brain also commanded the written-verbal abilities, since the left brain operates the right-hand side of the body. In a society in which reading and writing skills were paramount in education, social life and coping with technological advances, the left brain was necessarily dominant.

And We Cook Better, Too!

Along with these findings came the seemingly inevitable bigotry attendant upon new discoveries about people. The old prejudice against left-handedness now had some scientific backing to increase its stridency. The right hemisphere became the subject of a witch-hunt. The left hemisphere was lauded as the seat of all intellectual advances, while the right was increasingly regarded as unimportant except as the seat of mental illness!

Soon, the superiority of white males was said to be demonstrable because of their left hemispheric dominance, while those in whom the right hemisphere was deemed dominant were judged inferior. The list of these despised unfortunates included women, all non-white races and the lower classes.

Even today, the systems we use for discussion and discovery of issues tend to be geared toward left-brained organization and debate. The 'off-the-wall' right brain can be a startling and disruptive breath of fresh air. Philip D., a Left Double whom we interviewed on his vacation, told us this:

"I have the reputation of being the agent provocateur *at the firm. When we have our weekly meetings, I will often challenge the group's approach to an issue, perhaps even try to eliminate it altogether to move the meeting on. I like to have everyone's agendas out in the open, as well. This doesn't make me the most popular guy in the room, but there's so much dodging the solution for the sake of protocol, that I get tired of it. It helps, of course, that I'm the boss."*

One French researcher, Gaetan Delaunay, whose country had been defeated in the Franco-Prussian war, put all Germans on his list of inferior races. He then went on to describe how the people of other countries turned towards the left while performing their national dances, thus proving their inferiority to the French, who apparently turned to the right!

Such bigotry masquerading as scientific study spawned the general opinion that the left hemisphere was the human, intelligent, male, reasonable side of the brain, while the right was the animal, emotional, female, mad side. This was a reflection of the general thinking of the beginning of the twentieth century and served to reinforce existing prejudices with a 'scientific' foundation.

Women's Rights, Men's Lefts

So, the popular concept among the brain researchers of the nineteenth century was that the right brain was the seat of the non-intellectual, hysterical and weak traits in behavior. This happened to coincide with the attitude of the majority of men towards women. The societal stereotypes of 'womanly' behavior were strengthened, lending credence to the image of the over-emotional female by giving it the stamp of scientific research. This gave credence to men regarding women as dependents to be protected and indulged by them.

Thus, European men proved to themselves that they were destined to be the leaders of the world, and that all right-brained humans should realize their own inherent inferiority.

The Good, the Gauche and the Sinister

The prejudice against what were seen to be right-brained qualities at that time also extended to left-handed people. They were seen as evolutionary throwbacks by many and the common view was that left-handed children should be beaten, if necessary, into right-handedness. The words for *left* (especially *gauche* in French and *sinistra* in Latin) acquired negative connotations. There are also Biblical references to the people on God's right hand entering Heaven, while those on God's left hand are cast into Hell. So, being left-handed became synonymous with clumsiness, ineptitude, awkwardness, deviousness evil and misfortune.

Yet recent research has shown that left-handers tend to use more of their brain than right-handers. There are left-handers who have their verbal centers in the left cerebral hemisphere, like right-handers, yet have developed the right hemisphere's motor skills. There are left-handers whose verbal center is in the right hemisphere, which brings into play a right hemisphere skill most people do not have. Then there are those left-handers whose verbal skills are shared by both the left and right hemispheres. This is because most of us are not just left- or right-handed. There is a 'sliding scale' of handedness from strongly left to strongly right.

Dr. Marian Annett, at England's Leicester University, has theorized that there is a left-brain-bias gene, which is unique to humans. About 50 per cent of people inherit the gene from just one parent, which makes them mildly right-handed. If the gene is inherited from both parents, which occurs in about 30 per cent of the population, it causes strong right-handedness. The remaining 20 per cent do not receive the gene, in which case they have a 50 per cent chance of being either one. Thus we arrive at 10 per cent of the population being left-handed.

It seems that being right-handed may actually compromise visual and spatial skills in favor of language skills and that left-handers tend to be more adept than right-handers at learning mathematics, engineering and chemistry.

This would probably surprise those earlier researchers who derided left-handers, women and foreigners as being little higher on the evolutionary scale than chimpanzees.

Mind of Two Beings

While more recent researchers have managed to tone down the rhetoric in their findings, there is still argument as to what are the precise functions of the two hemispheres. In 1993, Ronald Puccetti, Ph.D., a scientific philosopher, posed the theory that there are not only two minds, but two separate **people** within the human brain.

It has long been known that the right brain, while usually incapable of speech, could transmit signals to the left brain when it was doing something wrong during an experiment.

A startling instance of this was the test given to a patient who had suffered cerebral injuries that had caused disconnection between the brain hemispheres. This split-brain condition confines sensory impulses to the hemisphere at which they first arrive. The exception is pain, which is sent to both hemispheres simultaneously, via the thalamus.

The patient, who was normally right-handed, was shown a group of objects which was then hidden from view. An object was chosen that had to be identified by his left hand. The verbal left brain had to guess what was in the hand, since it was not feeling it. The non-verbal right brain *was* feeling it, but could not speak its name. The left brain guessed wrong and misidentified the object, which was a pencil. The right brain, which knew what the object was, made the left hand turn the pencil and press its thumb hard onto the pointed end. When the left brain felt the pain it identified the pencil correctly as being the only pointed thing it had seen in the group of objects.

This showed the right brain using the kind of non-verbal reasoning that babies use when problem-solving before they can talk.

Puccetti theorized that the right brain is capable of being the dominant brain, but, since most people are right-handed, the left brain gets all the teaching and simply becomes more learned in the skills it develops. So, if the left side of the brain ceases to function, can the body which contains it be said to be the same person? The body would not be able to speak because the right brain never learned how to do so. Logically, Puccetti says, it cannot be said that the same person learned and did not learn, therefore the normal brain must be two psychological systems with different learning histories. This definition of the brain could be taken to mean that the brain actually is made up of two separate *people*, two separate cognitive consciousnesses, not just two hemispheres and one consciousness.

Our Left-Brained Culture

Leonard Shlain, M.D., in *The Alphabet Versus the Goddess: The Conflict Between Word and Image* (1998), theorized that the invention of the written word gave rise to the dominance in culture of the left hemisphere. That shift caused humanity to move away from peaceful reverence for the goddess/Earth/Gaia and caused the vengeful male god religions to flourish, with their disdain for the natural world as something humans were put here to exploit, even destroy if they pleased. His hope was that the global use of the Internet, with its mixture of word and image, would spur a resurgence of right-brain influence on humanity.

In his epic book on laterality, *The Master and His Emissary: The Divided Brain and the Making of the Western World* (2009), Iain McGilchrist explains in detail the differing modes of perception of the two hemispheres. Recent studies show that the two brains share many tasks, but the left brain's continuing dominance over the right brain bodes ill for our future. The lack of perception of connections between objects, patterns, people and ideas – seeing discrete parts instead of the whole – leads to lack of emotional involvement, lack of empathy, a concentration on utility without higher purpose, reductionism, even dishonesty.

Fascinating split-brain experiments show that when the left brain is forced to pick an object relating to what the right brain is seeing, it will pick a random object, then create a fabricated rationalization to try and explain the choice it has made. The left brain evades responsibility, behavior that is certainly endemic to the halls of (male) left-brained power in business, politics and war.

Being of Two Minds

The debate over the relationship of the left and right brains continues, but the idea that the two hemispheres of the brain govern different facets of the personality is now widely accepted. The left and right brain functions are now seen as complementary, normally working together in their own separate functions to form the behavioral patterns that make each of us unique human beings. These hemispheric differences have been shown to be present at birth, thus setting the stage for the baby's personality to form on the basis of its pre-determined hemispheric interaction.

The separate processes of the left and right brains are listed in the following table. The information was synthesized from several sources and reflects the work of leading researchers in the field, such as Robert Ornstein, Michael Gazzaniga, Anne Harrington, Sally Springer and Georg Deutsch. Simply put, the left and right sides of the brain process external and internal stimuli in different ways. Generally, the left brain is the more analytical and logical of the two, while the right is more holistic and intuitive. In the table you will find that there are qualities on both sides that you possess. Even if you have been previously told that you are either a right-brained or a left-brained person you will show preferences from both hemispheres. The balance of left and right brain influences creates an individual's unique personality profile.

Take a quick test drive of the BrainLines system:

• Read each trait.

> • Give **each trait** a score on a 1 to 5 scale, depending on how much it applies to you.
>
> •Remember, this is not an either/or test. For example, don't score *either* "Logical" *or* "Intuitive". Score all the traits.

• Total the Left-brain traits, then the Right-brain traits.

• Compare the scores of Left and Right.

> • If the Left comes out way ahead, you perceive yourself as Left-Dominant. If the Right is way ahead, you think you

are Right-Dominant. The closer they are, the more brain-Balanced.

LEFT BRAIN TRAITS	RIGHT BRAIN TRAITS
WORD-ORIENTED: finds learning from written instructions and information easier	IMAGE-ORIENTED: finds learning through visual aids easier, thinks of things in terms of images
LOGICAL: uses rational approach to reach a conclusion	INTUITIVE: follows hunches to an overall, open view
SYSTEMATIC: prefers to process information methodically, analyze step-by-step	HOLISTIC: sees information as groups of ideas, prefers overall views of situations
LINEAR: thinks sequentially, takes one problem at a time	RANDOM: thoughts may jump from one subject to another
STRUCTURED: concentrates on one activity, studying every detail and nuance. Connects elements in logical order	INFORMAL: divergent interests. Sees connections among many disparate elements
STABLE: likes to live life in an orderly way, likes to feel control over life	FLEXIBLE: likes to live moment-to-moment, spontaneously and keeping options open
CAUTIOUS: prefers to use proven procedures and accepted modes of thought, suspicious of change	ORIGINAL: likes to improvise, think of new ways to approach problems, new theories, new language

LITERAL: communicates better by writing and reading	SOCIAL: communicates better by talking than writing
INTELLECTUAL: considers all facts carefully before forming opinion	SENSORY: tends to act upon physical feelings and mental impressions
REALISTIC: prefers to confine thoughts to tangible subjects and pursuits	ARTISTIC: affinity for imagery, likes to experiment with sensory stimuli, shapes, colors
HANDY: has technical aptitude, affinity for tools, likes the logic of machinery	NON-MECHANICAL: not interested in understanding mechanical concepts
ORGANIZED: likes to categorize, put everything in order	FREE-FORM: prefers to let things fall into place
RESPONSIBLE: has no problem with repetitive tasks, works steadily towards goals without distraction	DISTRACTABLE: dislikes repetitive tasks, attention easily diverted to another interesting project
SERIOUS Does not 'get' metaphor or humor, deals only with perceived facts	HUMOROUS: Understands irony, metaphor and hidden, implied meanings

Where Am *I*?

The preceding table showed the preferences of the separate hemispheres of the brain. Nobody with a healthy brain exhibits purely one-sided preferences. We have some propensity towards one side or the other with an influence from the opposite side in a ratio that is unique to each of us. In order to think and react to exterior stimuli effectively we require a combination of both modes of cognition.

• Go to the mirror and check your BrainLines reading against the following faces.

• Does your perception of yourself add up to your reading?

BALANCED DOUBLE:- Two lines of equal length, one on each side of the area between the eyebrows. Balanced Doubles combine the qualities of the left and right brains in equal measure.

CENTER:- A single vertical line in the center of the area between the eyebrows. The line will sometimes be to one side of the center, but the lower end of the line does not turn to the left or the right. Centers can access both sides of the brain at once or either side, at will.

CLEAR:- No clearly-defined pattern of vertical lines on the area between the eyebrows. They may show a jumble of partial lines, dimples and swirls, or no markings at all. The Clear person feels influence from both sides of the brain.

LEFT CLEAR:- Single line on the left-hand side of the area between the eyebrows. The Left Clear person combines right brain and Clear influences, with the right brain being stronger.

LEFT DOUBLE:- Two lines, one on each side of the area between the eyebrows. The line on the left-hand side is longer than the one on the right-hand side. The Left Double person combines left and right brain influences with the right being the stronger.

RIGHT CLEAR:- Single line on the right-hand side of the area between the eyebrows. The Right Clear person combines left brain and Clear influences with the left brain being the stronger.

RIGHT DOUBLE:- Two lines, one on each side of the area between the eyebrows. The line on the right-hand side is longer than the one on the left-hand side. The Right Double person combines left and right brain influences with the left brain being the stronger.

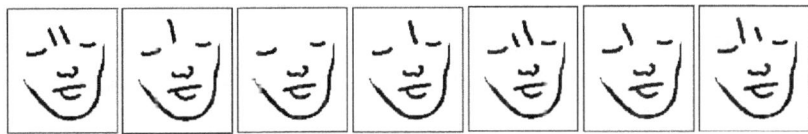

As a person with a Clear reading, I wonder how my friends with BrainLines can get through life when, for them, it seems to be a rollercoaster ride of exhilarating or painful experiences. I like my life to be as stable as I can make it and have it be as normal as possible. I spend enough time working, keeping fit, taking my kids to the game, saving for Christmas presents and paying the bills without worrying about stuff like the "meaning of life." Let me know when they've figured it out.

CHAPTER THREE
APPLYING THE KNOWLEDGE

That which God writes on thy forehead thou wilt come to.

The Koran

Humans have been endowed with many skills that have enabled us to be supreme survivors. We have physical adaptations, unique mental capacities and the cognitive processes that are divided between the left and right hemispheres of our brains. The diversity of all these adaptations has enabled the human race to exist and prosper in widely differing environments.

The brain's functions are so variable among individuals as to be unclassifiable into any more than a few general, basic types as an aid to insight into our basic mental profiles. Once we have gained that insight, we can explore further the effects that our life circumstances have had on our basic personalities. Acting on this information, we may change our lives accordingly to better suit our psychological propensities.

Go With Your Strengths

When we know our true strengths, we can tailor our lives accordingly. We do not have to put ourselves in situations to which we are totally unsuited. We do not need to push our weaker traits beyond safe limits. We can eliminate the danger of overtaxing our entire system and experiencing the pathological results of mental and physical stress.

This can occur in seemingly innocuous situations. There was the example we found Jerry, a proud father who had spent a lifetime building his own real estate business, a field admirably suited to his talents. Naturally, he wanted his son, Jerry Jr., to share in the joy and good fortune that he had experienced in his life. He did not realize that his son had different BrainLines from his, indicating dissimilar cognitive function and personality. The father pushed the son to take over where he left off. He tried to form his son in his own image, and could not understand why the son, intelligent and hard-working though he was, could not make a go of it. In fact, the son seemed unhappy and even ungrateful that he should be chosen as the successor, the proverbial chip-off-the-old-block.

We talked to both of them. Jerry Sr. told us, "I've worked my whole life to build this business and raise my family. It's been good to all of us and I can't understand why he wants to throw it all away."

Jerry Jr. said, "I've tried to tell Dad that I love him, but that I'm just not *him*! I wish I could do what he expects of me, but every minute I spend in this place seems like an hour. I feel like I'm wasting my life."

While there were the best of intentions on each side, both the Jerrys ended up bitter and unfulfilled. They were victims of the stress of the son being placed where he did not belong. There was an argument soon after our interview, the last straw for both of them. A year later, Jerry Jr. had started his own Web design company and was in his element.

"I'm on my own now...working all hours. This company's going to take off and Dad's going to be really proud of me."

All Play and No Work

Once you know where your strengths and limitations lie, once you accept that they exist, you can realize the potential of who you are. This may be at odds with how you visualize yourself. The knowledge of your true strengths gives you the opportunity for limitless achievement by using the best talents you possess.

Many people discount certain innate abilities and traits they possess that enable them to accomplish tasks with little effort. They may think of work as being the thing they do that takes time and effort, while something that we can do with our eyes closed cannot be classified as work, simply because it is too easy.

But there are those who realize that if they do what they do best and concentrate on using the innate talents they possess, they have the potential to create the kind of life for which they are genetically equipped. Such people are the leaders in their fields, those who are born to do what they do, those who have realized their potential by going with their strengths.

Diana always seems to do everything right. Although there wasn't much money in the family, she set her goals for an education, worked through college and earned her doctorate in music.

She was snapped up by an eastern university and began teaching and playing in a string quartet that became popular in the area. Later, she married the Chair of the Music Department and continued sailing along with seamless ease.

"I can only guess that my parents noticed when I was very little that I was constantly tapping my fingers and feet. I would also imitate any sound I heard. They saw my musical potential and set me on track. Amazing, because neither of them has much interest in music."

Athletes, soldiers, writers, scientists, carpenters, farmers, philosophers, artists, truck drivers, all have their obstacles to overcome, their challenges to meet and their benefits to reap from what they do. If they choose their profession well, they will find the pleasures outweigh the pain and they will experience the feeling of being in the right place at the right time. Learning to trust and make friends with your inner guidance system makes it easy to find the best outlet for your own creativity.

Creative Juice

A few words about creativity. Many of us think of it as being applicable only to artistic pursuits. It was a common response from subjects of BrainLines research, when they were told they had creative traits, that they "couldn't even draw a straight line." But creativity is not exclusive to certain professions, pastimes or products, nor even to the right brain.

In all walks of life, in day-to-day living, there are opportunities for creative activity and thinking. Creativity includes ingenuity, originality and inventiveness. Do you know someone who can work magic with food or fabrics or wood, someone who can always find an answer to people's problems, someone who has the gift of the gab, someone who can always make you laugh?

Creativity is a special perceptiveness, an ability to combine various concepts to produce new ideas and things. Right-brain creativity is a function of abstract vision, the kind of artistic sense that is most often regarded as creative talent. It is often highly improvisational.

Left-brain creativity operates with deliberateness and less guesswork. There can still be flashes of inspiration, which will come after the research has been done. The left brain is responsible for much that is produced in industry; houses, bridges, cars, etc. The left brain's process of putting together various disparate design elements is the creative act.

Most often, creativity draws from both sides of the brain in varying degrees. Even the most highly-engineered products of a left-brained can appeal to the eye as well. It is this kind of integrated creativity that is seen in the great works of art and science as well as every day life. It is a combination of inspiration and hard work, daring and caution, fantasy and reality, with tests, judgments and comparisons to follow. It is a mix of spontaneity and experience. What changes from person to person is the dominance of the left or right brain in the process.

Our dear friend Alicia, a Left Clear, has made a successful business of creating personal altars. By combining her preferences of color, texture, spatial sense, structure, strategy and detail, she is developing an increasing market for her unique creations.

The spiritual aspect of personal altars, enclosing images and objects that have special meaning to her and the client, is a natural for Alicia. She has spent a good deal of her life in pursuit of inner knowledge. Now, in her mid-thirties, she says,

"I feel very lucky that I've been able to translate my spiritual visions into a commercial livelihood. I'm a very happy person."

Another Brick in the Wall

Many people find themselves in careers or relationships to which they are psychologically unsuited. They may feel compelled to this by external forces such as the cultural pressure to be "better off" than one's parents or to be married and have children by a certain age. Sooner or later they succumb to the internal feelings of lack of fulfillment, anxiety, alienation, of being trapped and stressed to the limit. Asking ourselves specific questions about life choices and answering them truthfully, stating our innate preferences, can ultimately direct us towards a healthy, enjoyable and truly successful life.

When using the word "successful", we are hamstrung by the meaning that it has been given in our century. Success has come to be measured solely by the amount of income that can be generated by one's endeavors, whereas that used to be just one of the possible criteria.

When success is judged by one's credit rating, rather than the fulfillment achieved from work, relationships and life as a whole, the pressure is immense to enter a lucrative field, rather than a compatible one. Thus we are confronted by the image of the high-school senior who, on being asked what he plans to do with his life, answers, "Oh, I want to be a doctor or a lawyer." This answer shows a commitment, not to either of these honorable professions in themselves, but only to the high income they command.

The Doctor Will See You Now, Mr. Down-Payment

Certainly, there is nothing wrong with earning the fruits of one's labors. But if you were going to see a doctor or lawyer for help with a life-changing problem, you would prefer them to feel some kind of calling to their work, and hence your problem, rather than their seeing you as this month's mortgage installment.

Stress-related illness has also become the norm, and little wonder when people shoe-horn their psyches into jobs that may be an unsuitable type of work for them. They become trapped by the search for ever-greater income and, once it is attained, the fear of losing it. This is the death of the creative in us. Hollywood, for example, is full of ex-visionaries who started out with grand creations in mind, only to be mired in the mundane for fear of rocking the boat and losing their financial cushion. The pertinent question is, "Can I downsize my expensive tastes for the sake of fulfilling my potential creativity?"

The bottom line is, no matter what endeavor you choose, if it is the right one for you, there is a greater chance of being good at what you do. You increase the possibility of feeling happy doing it, getting greater fulfillment, feeling less stress.

Critical Mass

At one time or another all of us have criticized ourselves for coming up short in a situation that demanded more than we were equipped to give. We have all experienced flashes of envy when we see someone doing something we want to do, but better than we could ever do it. Inside, we have thought less of ourselves. Hence, we may give up trying to achieve something because of fear of failure. If, at those moments, we could stop maligning our lack of skill or intelligence and realize that we are actually picturing ourselves as someone else, it would become clear that the burden we are placing on ourselves is unrealistic.

When we learn to take the energy that is wasted on trying to cope with situations that are moving us away from our strengths, and use it to find ways of experiencing more fulfillment, we will stop *tolerating* our lives and begin enjoying them.

Child's Play

As if our own expectations weren't enough to deal with, there are also the social norms. The need for categorization runs deep in our culture. For example, the pressures to engage in "manly" pursuits or maintain "womanly" ideals promote the kinds of character distortion that can lead people to live their lives in ways that have little to do with their own needs or desires.

Traditional roles for men and women can create burdens on boys who happen to be more right-brained and girls who are more left-brained than the "ideal" stereotype. Those boys who are more intuitive and sensitive than the cultural norm either have to try to mask their character or be considered "less than male". Likewise, girls who are more assertive than passive, more intellectual than sensual are judged to be "less than" the desirable norm for women in a patriarchal society.

As children, we pick role models with whom we can identify and compare ourselves. This emulation of those we admire can give rise to healthy learning examples and strengthening of talents through trial and error. Unfortunately, these experiences can be less healthy when they are the product of role models being chosen for us *by others*. Under these circumstances our lives may become as fictional as a cartoon, where we will never be the hero and in which there is no part for the real "us" to play.

When a parent or teacher can read a child's BrainLines and get an idea of the psychological foundation on which they are building, they can adapt their teachings for the ultimate good of the child. When children feel at ease with their traits and the expectations placed upon them are within their abilities, there is naturally less for them to fear in the learning process. Learning to prepare for life is an arduous enough task without having to do it twice; once for who you have been told you should be and once for yourself.

BrainLines and the Actual Self

So we have explored some of the problems associated with not knowing our inner selves. What if you are already suffering the effects of a life led in opposition to your innate attributes? How can the BrainLines system help?

BrainLines indicate your essential inner nature, the hereditary raw material of the personality. Your genetic foundations of talents, inclinations, needs and temperamental balance are potentials, not actualizations. They are the stuff of which personality is made, they shape your unique behavior patterns, but can cause problems if they are not used.

Culture, family, environment and education all play a part in the processing of the raw material. The BrainLines profile with which you are born may be suppressed by these strong external forces. Some innate propensities may be deemed so unsuitable that the suppression may result from your own learned fear of them. This can happen when a child tries to drown out impulses of which it has been taught to disapprove. Fear of exposure and humiliation are strong forces compared with innate tendencies.

An illustration of this is one of the BrainLines research subjects we met in California. Stan is a strapping six-foot-four product of the Southwest, with a deep Center BrainLine.

Stan felt a strong interest in and sensitivity towards the people he encountered from an early age. He was able to perceive their needs and was able to counsel them in a sensitive and intuitive way. In his teens, he was a trusted confidant to his friends in the church he attended. The minister recognized this and encouraged Stan to pursue a future in some kind of counseling, perhaps the ministry.

Although this appealed to him, he succumbed as a young man to the accepted view of intuitiveness and sensitivity as being too "sissified".

"If my buddies on the team had known, they'd have given me a real hard time."

His father encouraged him to study structural engineering. Stan had the aptitudes and liked the idea of designing what he saw as sculptural structures and his dad, a carpenter, thought it would be "a step up." By the time he finished college he realized that the constraints of the field did not appeal to him, but he had invested too much time and money to try something else. Besides, there was a good living to be made in engineering. "And my Dad would've cut me off if I'd told him I wasn't that interested any more."

By middle-age, Stan had become an adequate engineer. "The job was just O.K., but my heart wasn't in it because there weren't enough challenges."

He worked long hours, trying to compensate for his low self-esteem and maintain his lifestyle. He had no time to make real friends. In trying to ease the pain of loneliness, he became addicted to alcohol, which made impossible any real interaction with anyone. It finally cost him his job. When we met him, he was living in San Diego and had luckily found himself a counseling position at a rehab clinic. Stan was finally fulfilling his spiritual needs.

"I can't believe I let so much of my life slip away, doing something I didn't really like. I guess finding out the hard way was the only way I could make the change. For the first time in a long while, I feel good inside and I like to think I'm making a difference in people's lives. I'm seeing a lot of guys on the bottom, where I was, and their stories aren't too different to mine. It's all about feeding your soul."

Our basic potentials do not disappear. The energy required in suppressing them causes an unconscious fatigue on the psyche. This can lead to disorders, feelings of malaise, of being valueless to society, of lack of fulfillment. An unused personality trait will act like an atrophied organ, becoming a center for dis-ease and diminishing the efficiency of the whole organism.

The remedy for such a situation is self-actualization, a process of re-discovery of the innate self, leading to new modes of perception and action. BrainLines can be helpful in acting as a catalyst and guide to your potentialities. With a map of your inner self, you can be less afraid of the unknown. The greatest human fear is change. Any step forward means leaving the familiar and the comfortable behind. But if you see your potentialities as positive gifts, you are not "giving up" your weakness, but gaining your strengths.

Self-actualization is the demolition of a false construct or representation of your Self, and the replacement of it with a new design using your own raw materials. Making the true Self real, or actual, is an important step towards being able to live a spiritually fulfilled and self-expressive life.

As the process continues, the division between yourself and everything else becomes less distinct. Going inside yourself does not mean alienation or loneliness, but an acceptance of yourself as a part of the whole. Still a unique being, but going beyond notions of selfishness and self-consciousness to self-inclusion. This sounds sublime, but self-actualization does not remove the everyday hurdles in life. It teaches you how to use what you were born with as efficiently as possible to meet day-to-day problems. That is when the value of the process shows itself. Working out difficulties becomes easier and fun when you work with your own innate gifts. Plus, your sense of humor will improve!

As a preliminary step on the journey to self-actualization, the BrainLines profiles are guides to the possibilities of each of the seven broad combinations of left- and right-brain strengths. Once you are set on the road, it becomes your journey. You get to choose your direction, depending on how you wish to use those strengths.

Do not be disappointed if the profile of your BrainLines reading does not describe you 100%. Your profile is an illustration of your potential, not a narrow category that you have to match in every detail. You are on your own path, because on this pin-prick of reflected light in our solar system, the only category that matters is that of "Earth inhabitant." Beyond that, we are separate and unique beings.

PERSONALITY PROFILE FOR
BALANCED DOUBLE

The Balanced Double *displays two lines of equal length, one on each side of the forehead between the eyebrows.*

Quick Summary for *Balanced Double*

Read this to get a feel for what BrainLines can tell you about yourself and people you meet. The Full Profile begins on the following page.

Scattered? Sure. Frenetic? Maybe. Multi-talented? You bet! You have an idea-a-minute, from the abstract poetic to the hard edge of sequential logic. All options are open to you, which makes you a great problem-solver, if you can focus long enough. Some of us can't see all sides as you do. As you go out in the field, sniffing every scent, finding out all the information, others will be following the footpath. Both may reach the same point, but you will have discovered so much more along the way.

Relationship tip: You're juiced by happiness, and will bend over backwards to stay that way. Such flexibility is like money in the bank when it comes to a relationship. Yet you don't wimp out when times are tough. You find every possible solution. When you choose the perfect partner, you will be an inspiration to them with your whirlwind of ideas and interests. Two *Balanced Doubles* together? Anything is possible!

FULL PROFILE FOR *BALANCED DOUBLE*

The Hard-Wired Brain

The people with a *Balanced Double* reading are interested in many things, from the technical to the spiritual, since they possess the qualities of both the left and right brains equally. These people are able to see with overall vision and attention to detail in creative and organizational ways. They have the potential of both verbal and visual skills that will support their innovative drive.

They are full of information and are not usually shy about sharing their knowledge. Sometimes, their ideas come so thick and fast from both sides of the brain that it requires a conscious effort to put their images into words. Once the issue they are dealing with is thought through, they can use their balanced traits to make their ideas accessible to their audience. This quality makes *Balanced Doubles* gifted leaders and co-workers who can bring excitement and practical creativity to the workplace.

Balanced Doubles are fully aware of the sensory and emotional input from the world around them. They have the uncanny knack of receiving this constant input, sifting it and balancing the attention of both sides of the brain to each part. They are drawn to every facet of life and the symmetry of their left and right brain processes makes this easy for them, enabling them to find the middle road between multiple and sometimes conflicting interests.

Balanced Doubles are best at combining insight and judgment to come up with original ideas and plans for the future. They can see the connections between all the information they receive and put together a plan combining all the elements, using practicality and visionary skill.

Listening to a *Balanced Double* going through the process of thinking out a new plan can be intriguing for other people. *Balanced Doubles* can string together images that jump seamlessly from one seemingly unconnected subject to another.

Feeling the Feelings

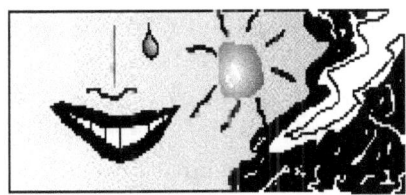 All readings with BrainLines denote sensitivity to others' emotional states and *Balanced Doubles* are full of this quality. They are always willing to give advice on any personal subject. Finding new ways of looking at old problems is their greatest strength, so the perennial dilemmas of the human condition give them plenty of interesting material with which to work.

The broad sweep of their vision is especially useful in relationship problems, which more often arise from troublesome details than from total breakdown of communication. *Balanced Doubles* can look at such situations from all sides, from inside and out.

Relationships, friendships and partnerships of all kinds are very important to these people. They will be concerned over emotional distress in their companions and family. *Balanced Doubles* are possessed by their interest in what makes people tick and how life can be improved for all. They always have ideas and emotional concerns on this score and try to convey them in language suited to the individual with whom they share them.

When *Balanced Doubles* are involved in activities that interest them, they will bring to them a passion that will inspire some and distract others. While they are aware of other people's states of mind, *Balanced Doubles* may get so emotionally involved in their interests that they become somewhat blinded by their own power and their effect on others. When they stop to think about this, *Balanced Doubles* will be disturbed, since they have the sensitivity that makes them want the best for everyone. The combination of involvement in their work and concern for others, when balanced out, makes them effective leaders, who will build long-lasting working relationships.

Social Ability

Complacency, boredom and shutting down of emotions have nothing to do with *Balanced Doubles'* lives. They make good counselors, staying detached, watching from the outside and taking interest in the interplay of emotion. They can work around discord because they see it as an inevitable part of human nature. Indeed, their own way of processing *is* a kind of internal discord. *Balanced Doubles* are able to separate themselves from the issues, consider them and disagree with the opinions of others without bearing grudges. They can appreciate that there is more than one side to every issue and that each deserves a hearing.

All the input they receive can make *Balanced Doubles* seem almost desperate in their need to manifest their original opinions and methods for dealing with situations. There is so much in life that interests them that they feel as if they have to get on to the next thing as soon as possible, in order to keep the creative juices flowing. If they can harness this energy and focus on the present, they will make excellent organizers and administrators, especially in community projects, who will instill their fellow workers with their enthusiasm.

Balanced Doubles, although they have the need to put their ideas into action, do not crave power and will share leadership, if they feel it is the better way to go. They will be open to being a part of the whole, an element in the structure, if the others are willing to share the load and have a similar vision of their goals. Tedium in work or social situations is unacceptable to the *Balanced Double*.

They have so much going on in their minds that there is always something new to draw their interest away from the boring. Sharing the leadership role may be evidence of an unconscious recognition that they will need someone to whom they can hand over a project, if boredom should set in.

When a *Balanced Double* enters politics, it will often be through the inspiration of working towards the common good. They will try to maintain this concern for their constituents even in a political system that tends to stifle altruism. Their concern will always be the improvement of a situation that they see as unsuitable. Attachment to a certain political party will be the key to getting their projects accomplished, rather than being a simple devotion to a structure within which they can hide. As politicians or activists, these movers and shakers will devote themselves to community challenges, perhaps at the cost of their own well-being and political career.

Those *Balanced Doubles* that enter community activities will be unselfish in sharing their time and energy and will commit to something they believe in, finding reward in contributing to the improvement of people's lives.

Dating and Relating

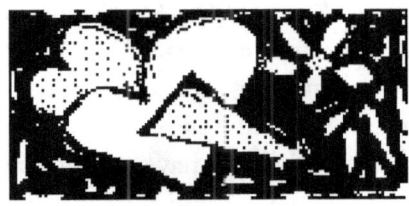

The ability of *Balanced Doubles* to view situations from both logical and intuitive viewpoints makes them very adaptable in working for the survival and liveliness of a relationship. *Balanced Doubles* will always look for the middle road in a relationship's problems. They are willing to compromise and suit their actions to their partner's in the search for harmony. This does not mean that they are weak-willed or uncommitted. They have their opinions and methods of dealing with situations, but will adapt, just as long as their partner shows the same willingness to work problems out. The *Balanced Double* knows that there are always viable options and don't get stuck on one course of action.

Their ability to disagree with opinions rather than with the individual means that they will not bear a grudge, even though they may argue passionately for their side of the issue. *Balanced Doubles* will play an active role in ensuring the longevity of their partnership, marriage and family. Their contributions to a relationship will often be designed to shake things up a little, to keep the interest alive and to avoid stagnancy and complacency. That is the key to any endeavor in which they are involved, if they are to stay with it.

They may sometimes seem inattentive and unconcerned with the present because they have so much going on in their minds that only they can connect. Sometimes it is better just to let them go off on their own mental tangent and be around when they get back.

Balanced Doubles do not picture themselves as willing participants in structured situations, but they just may come up with a concept that will improve a relationship or family life for everyone involved, and it is new concepts that excite them.

The flexibility of being able to access both sides of the brain simultaneously allows them any number of relationship options, but their first choice might be another *Balanced Double*, because of the possibility of sharing collaborative interests. Such a creative partnership would be fulfilling for them and their collaboration would require that they spend a great deal of time together.

Balanced Doubles would probably have such an understanding of each other's approach to work and romance that they might have a better chance than most of pulling it off. They would both resist allowing any part of their lives taking precedence over the rest. They would, however, need a full commitment to the concept of total collaboration to bring as much focus as *Balanced Doubles* can ever achieve to all aspects of the partnership.

If the partner of choice were a *Right Double* or a *Left Double*, the relationship would still benefit from the shared flexibility on all sides. Both of these readings would bring the same possibilities of creative collaboration as would another *Balanced Double*. The emphasis would be slightly more polarized towards the dominant brain function, but there would be an understanding among all three readings of each other's reactions and interests.

A *Clear* or *Right Clear* partner would be compatible, but levels of emotional intimacy would differ greatly. *Balanced Doubles* would be able to appreciate the traits of these readings, but would demand the emotional sustenance they need. They would be able to help *Clear* and *Right Clear* partners to access deeper feelings, enabling them to find methods of expressing their emotions more suitable to their pragmatic natures.

An interesting and challenging choice would be the adaptable *Center*, who would completely understand how the *Balanced Double's* mind works. *Balanced Doubles* would be challenged by overcoming the *Center's* detachment and reaching their level of vision and intuition. These lofty objectives would provide a focus and an inspiration to work with while trying to achieve their own goals. *Centers* may be cooler in their shows of affection, but their compatibility with the *Balanced Doubles'* perceptions of life would bring them close.

The relationship with a *Balanced Double* will never be a stagnant one, if they can help it. They look on change as a means of achieving stability in a relationship, variety as the spice of life and the dynamic force in a marriage. Partners should expect the unexpected from *Balanced Doubles*. They will continually look for ways to get a rise out of their partners, harmless and humorous ways of making sure that neither of them gets bored.

Travel and the satisfaction of their wide-ranging interests will play major parts in their search for marital contentment. Their partners should always be ready for a change of plans that have to do with the *Balanced Doubles'* passions in life.

Excitement, challenge and, above all, flexibility are the key to a relationship with a *Balanced Double*.

Compatibilities

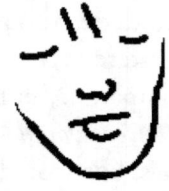

What are you looking for in a romantic relationship? Here are the major qualities to expect in a partnership when a **Balanced Double** pairs up with one of the other BrainLines readings.

Balanced Double with Center
Intellectual and emotional challenge.
Stabilizing influence of the *Center*.
Similar traits, but different styles.

Balanced Double with Balanced Double
Creative collaboration would be stimulating.
Emotional support and compatible traits lead to shared and equal responsibility.

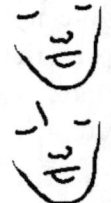

Balanced Double with Clear or Right Clear
Lack of intimacy without deep emotional involvement from *Clear* and *Right Clear*.
Plenty of space and time for *Balanced Double's* interests.
Stability and predictable relationship.
Compromise in style necessary from both partners.

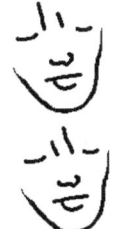

Balanced Double with Right Double or Left Double
Shared flexibility of thought.
Supportive traits, added creativity from both *Right Double* and *Left Double*.
Complimentary abilities and methods.
Romantic relationship is possible.

Balanced Double with Left Clear
Emotional rapport, combined with *Left Clear's* pragmatism. Calm creativity from *Left Clear* complements *Balanced Double's* frenetic style.
Long-term commitment possible.

Work That Works

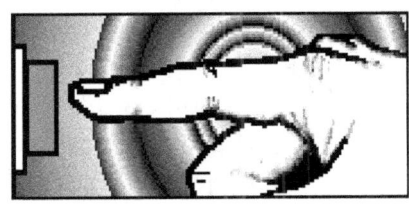

Balanced Doubles have the diverse talents of both sides of the brain at their instant disposal. This gives them eclectic tastes and views, tends to scatter their attention and makes them versatile and flexible.

What they bring to the workplace is a unique way of processing information and using it to come up with innovative ideas. There is apparently no limit to their vision, once they are interested in a project. They are the archetypal lateral thinkers, who will find connections and possibilities that no-one else would even consider, taking seemingly disparate concepts and linking them in new ways.

Their ability to combine practicality and imagination means that their stream-of-consciousness style of processing can produce truly sound and realistic long-term decisions. While some of their ideas may seem too innovative for the present, most of them will be workable solutions.

Balanced Doubles are open to any and all suggestions and each one will set off a whole new train of thought. They are full-time inventors and conceptualizers whose 'off switch' can be hard to find. They need to be given a fixed goal towards which they can work from any direction. As long as the goal remains constant the *Balanced Double's* long-term inventiveness can be limited to the current project. Given a free hand and an open-ended project, there is no telling where a *Balanced Double* could lead us.

When it comes to being a member of a team, *Balanced Doubles* will be a creative force, spinning off ideas, but parceling out the tasks to team members with different skills. In this way, everyone on the team can feel like a constructive force on the project. *Balanced Doubles* will try to lead others, depending on their rapid-fire delivery to get their ideas across forcefully. But they will also listen to what the other people have to say about their part of the project. They are always ready to listen to input that can cause a whole new set of tactics to emerge.

They will probably expect from their co-workers the same devotion that they feel for the work and may think it strange that everyone is not as completely enthralled in the project as they are. If they sense a lack of enthusiasm, they will try to consider the other person's point of view and will be willing to make changes in their approach, if they feel it is more acceptable.

Balanced Doubles value the input they get from other people as an aid to their efforts and are good at delegating responsibility or finding expert help as part of putting their ideas into action. They need other people's views to give a little perspective to their own ideas and will readily adapt their plans, since they know from their own experience that there are usually multiple solutions to most problems. Indeed, in the process of forming a course of action, *Balanced Doubles* will often start with the solution they come up with and work backwards from it, constructing the course of action in reverse. Their ability to look at situations from many angles enables *Balanced Doubles* to find matching concepts where others may only see conflicting information.

That kind of flexibility can make the people whose advice is being sought feel good about the work and themselves. The down side is that other team-members may have been concentrating on one course of action, only to have the course changed, which can cause confrontation. Those people's objections should be heard, so that the *Balanced Double* can integrate the input into the latest plan. It needs to be made clear to everyone involved that there is an open forum of ideas, that the goal is important, but that everyone will have equal input on the approach.

Conflicts may arise over differences in work styles, especially when someone with another reading wants to hang on to their idea. *Balanced Doubles* will typically look on conflict as another piece of the puzzle to be worked out. Because they are so flexible, they will have little attachment to their earlier decisions, taking it for granted that there will be discussion and confrontation in life and that it is all grist for the mill. They will discuss issues rather than people, thereby keeping interpersonal tension to a minimum.

They need to be given the freedom to use their strengths with as little restriction as possible. To them, rules are not meant to be boundaries, but parts of the equation to be ignored or changed if they interrupt the flow of possibilities. This may be seen as dangerously iconoclastic in a structured organization, but it is the only way for *Balanced Doubles* to access their full ability, which is a powerful force to have at one's command.

Balanced Doubles, with their varied interests, are suitable for many fields, especially those in which their creativity can flourish, the work is not repetitive, they feel stimulated and where there is social or working contact with others. These are adaptable and inventive people, equipped to work in any field, who are constantly pulled by their opposing but complementary traits, open to input on emotional and sensory levels and who will devote all of their considerable skills to work in which they feel free enough to operate fully.

BrainLines of the Rich and Famous
Balanced Double

Edward Abbey, Woody Allen, Maya Angelou, Yasser Arafat, Kristin Asbjørnsen, Richard Avedon, Matthew Barney, Ruth Bernhard, Jorge Luis Borges, Louise Bourgeois, Adrien Brody, George W. Bush, George Bush, Barbara Bush, Hillary Rodham Clinton, George Clooney, Daniel Craig, Julie Delpy, Dr. Dean Edell, Ramblin' Jack Elliott, Geraldine Ferraro, Al Gore, Katherine Graham, Alan Greenspan, Andrew Grove, Carla Gugino, George Harrison, Vaclav Havel, General Claudia Kennedy, Maggie Kuhn, George Marshall, Sean Penn, Vanessa Redgrave, Franklin D. Roosevelt, Sam Shepard, John Sununu, John Waters, Rachel Weisz.

PERSONALITY PROFILE FOR *CENTER*

The Center *displays a single vertical line of any depth or length in the center of the forehead between the eyebrows. The line will sometimes be slightly to one side of the center, but the lower end of the line does not turn to the left or the right.*

Quick Summary for *Center*

Read this to get a feel for what BrainLines can tell you about yourself and people you meet. The Full Profile begins on the following page.

Did anyone ever tell you they thought you were reading their mind? Have you ever felt like you knew the answers to questions before they were asked? No surprise. You are equipped with a DSL connection to your intuition, which means your hunches sometimes surprise even you. You've also got a knack for finding a logical way around life's challenges. You can do it all, if you can let go of the reins on your mind and trust your instincts.

Relationship tip: You're the most interesting person you know, and you can take love or leave it. But once you find the person who can keep you stimulated on all levels, you can dig deep into your reserves of love and compassion. You'll need your alone time, but don't overdo it. Your partner will probably want more shows of affection than you anticipate. Give it a whirl, you'll be making both your lives all the richer.

FULL PROFILE FOR *CENTER*

The Hard-Wired Brain

 People with the *Center* reading have a unique vision of the world around them, thanks to their ability to access both sides of the brain at will (with no dizziness). They have a complete lack of boundaries in their outlook on the world and their approach to the possibilities it holds for their lives. No stopping them!

They have the panoramic view of life that is analogous to looking down on a house without ceilings or roof. They can see the relationship of every component to the totality. Everything is joined to the universal whole. They have the ability to study the details of all the parts that go to make up the big picture.

Centers have verbal skills along with being orderly, painstaking, logical and focused. They are entirely comfortable with spiritual, religious, philosophical and metaphysical matters of all kinds, using their grasp of abstract notions to convert concepts into powerful internal imagery.

Centers are intuitive to the highest degree and those who have learned to use their intuition will be known as visionaries. They have the intuitive powers of both sides of the brain: instantaneous hunches from nowhere; coincidences of 'knowing' where to be and what to say; flashes of inspiration that strike after learning everything about a problem and sleeping on it; the foresight that comes from their long-range view. Once they learn to trust all their 'voices', *Centers* have extraordinary powers and depth of feeling.

They are aware of a constant flow of outside information through their senses. They can absorb this information and let it flow through them as they sift it and use it as sustenance for their rich, complex inner life, without breaking their concentration on present pursuits.

Centers are never at a loss for words, nor do they waste them. They use their verbal skills in a calm, reasoning way, explaining their processes and conclusions, and can give forceful arguments for their views and decisions. They believe that every problem has a solution that can be reached through investigation and the evaluation of alternatives, both through intuition and logical assessment. They can be made to change their minds on one of their own decisions by well-considered argument of an opposing viewpoint, reinforced by detailed data to support its conclusions. Once they are convinced that a new course of action is called for, they will devote their considerable powers of concentration to it.

If *Centers* do not realize or are discouraged from using their great potential, they have the highest risk of suffering from psychic chaos and lack of self-respect. Choking off their talents is likely to make them indulge in self-destructive behavior, such as alcoholism and drug abuse.

Feeling the Feelings

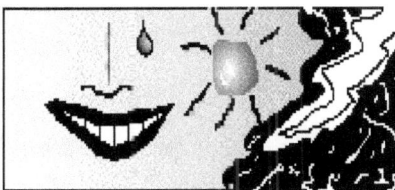

Centers typically have a well-developed sense of humor, which feeds on the comparison between their holistic vision of life as a vast interconnected pattern and other people's attitudes and hang-ups about the minutiae of life. *Centers* see through the pomposity and arrogance of self-important people (especially themselves, if they find themselves slipping into that sort of behavior) and can laugh at it. Their humor may tend to be dark because there seems to be so much that is laughable in the worst of human nature when seen in a universal context. The cynicism of that kind of humor is the *Centers'* way of quelling their own fears and seeing past the misery and mischief in life to their own sweeping view of the past and future.

Centers who are given free rein to use their logical and intuitive talents will have great faith in their own judgment, which, along with their detached manner, may seem like arrogance. Behind that confident shell, however, is a sensitive instrument, processing and filing information at all times. The detachment is the cushion between them and what can be a painfully stimulating world.

Social Ability

Despite their best efforts to block it out, there is a deep sensitivity to the *Centers'* dealings with people. It forms a psychic link that helps them to maintain the affection of co-workers and family members whose emotions are closer to the surface. *Centers* can often surprise themselves by their sympathetic reactions to others' troubles and will show the depth of their concern by offering rational solutions to them. There is a compassion present in them that may not show in the *way* they say things, only in *what* they say. Such nuances may be missed at the time by those in distress, who may expect a more overt show of warmth than a *Center* can provide.

This makes *Centers* sound cold, but that impression would ignore their deep awareness of other people's feelings. Their intense empathy would overload them emotionally if they could not separate themselves enough from other people's pain and confusion to get the larger picture of the problem. With that kind of overview and understanding of the other person's emotional needs, they can offer original solutions to problems that will serve the person best in the long run and makes *Centers* well-suited to healing, counseling and spiritual work.

Their ability to concentrate on the problem at hand makes them efficient thinkers, simplifying complex problems by filtering out extraneous details and zeroing in on the items that really matter. For *Centers* this is the best and most efficient way of getting on with life with as little trouble as possible.

Their desire to make life run smoothly for other people attracts *Centers* to community and charitable work. Their qualities of initiative, guidance and practical compassion will spur them on to work for the common good. That efficiency in the service of others is more important to the typical *Center* than the actual power and prosperity that a public service career can bring. While they will not reject the rewards of their efforts, they will try to do work that they feel justifies such rewards. This caring attitude will be reflected in their work and in the respect and admiration that *Centers* often command.

Because *Centers* can see the big picture they have a wide tolerance for the ideas and actions of others. They are flexible in their views and realize that all our endeavors are temporary in the long run. This may make certain careers short-lived for a *Center*. Strict conformity to a party credo or corporate agenda at the expense of their own opinions would go against the grain for a fully-functioning *Center*. That sort of blind loyalty and idealism, or the appearance of it, would seem senseless to them.

No matter what they do, they will be respected for their broad-mindedness, ethical convictions, progressive thinking, visionary concepts and for their perseverance in getting the job done.

That is why so many leaders in business and government are *Centers*. Their reasoned judgments and quick decisions, combined with their high principles and impartiality towards others, win the respect of those with whom they work. They are cool in stressful conditions, concerned with the welfare of their subordinates, adept at avoiding needless risk and focused on the goal.

These are charismatic, original thinkers with hidden emotional depths, who like to make life run smoothly for themselves and others, yet with the detachment of those whose customary view is of the big picture.

Dating and Relating

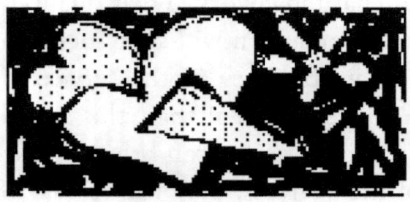

The partner of a *Center* has to come to terms with the fact that *Centers*, while capable of deep levels of love and devotion, are less dependent on partnership. *Centers* have such a rich and complex inner life and their sensory systems are so acute, that they are likely to be pre-occupied with their thoughts and feelings. At every waking moment, *Centers* are dealing with everyday life while simultaneously paying inner attention to their abundant sensory input and the images that arise from them. Their busy brain sometimes gives them an air of distraction, of not being quite 'present'. This usually happens during their relaxation time at home, when the *Center* has gone 'inside' himself, away from work and the world. It may be in these rare moments of free time that a partner will want to discuss something. The partner may have to wait until the interior sorting process is done.

The ability of *Centers* to focus and their desire for intellectual and spiritual growth require that they select a partner who will not try to deter them in these quests and can accept their busy schedule without making too many demands on them. Of all the readings, they are the most emotionally self-sufficient and, while they can sense the feelings of those around them, they will not be effusive in their show of caring. Their strength lies in their sensitivity to others, the originality of their clear and ordered solutions to problems and their calm attention to the needs of people around them. They have the capability of handling any emotional demands placed on them when they have time.

A good choice in a spouse would be another *Center*, with the understanding that their respective careers and interests will take much of their time. Alone time with each other might have to be scheduled. Two *Centers* would have to realize that emotional ties will always be a distraction from their work, and accept the situation.

They have the unique ability to adapt to any of the other BrainLines profiles, but on their own terms. The combination of their strength of will and holistic way of functioning can have a bulldozer effect. Partners must be prepared to stand their own ground on issues that are truly important to them, or sacrifice their own needs. Both will require deep understanding to make the relationship work in ways that are good for them both.

The partners of *Centers* need to allow them time alone, for meditation, daydreaming or free association of their thoughts, all of which are synonyms for 'input processing'. It is important to them to have time in which external stimuli are reduced to a minimum while the sensory system relaxes and the memory files are sorted.

Centers primarily see things and people in terms of their 'vibrational energy', their intensity of purpose. They have high standards that they expect to be met by their friends. They want to be challenged and stimulated on all levels. Partners would have to get used to this approach, trusting in the *Centers'* intuitive and harmonizing abilities to see them through.

One advantage of dealing with a *Center* partner is their flexibility of thought and willingness to compromise on issues that do not pose a direct threat to them. While they may seem intransigent in some ways, they are unlikely to become polarized, stuck on one side of any such issue. Part of this comes from their ability to see humor in every situation, especially fixation of ideas. Be prepared for unexpected laughter in times of crisis.

Centers are so in tune with their own feelings that communication should be easy for them. They have no fear of vulnerability through self-disclosure, so are able to open themselves completely to the idea of compromise and mediation. They can be good listeners, empathetic and original in response to problems and have deep wells of love, compassion and devotion for the rare partner with whom they will share everything they are.

Their affection, which may not be expressed as often as some partners might like, will be shown in how they try to make life run smoothly for everyone involved, with decisions based on lasting needs rather than whims of the moment. Their style is to give their partners plenty of space and not engage in too much small-talk. This translates, in some people's minds, to a lack of affection, but a typical *Center* will see it as giving their partner the freedom for self-determination and expression that they themselves desire. This may be taken to extremes, in which case *Centers* need to be reminded that they should devote more effort to nurturing the relationship.

In a *Center's* mind, any relationship that has been compelling enough to interest them in the first place is, by default, running smoothly. Once the practical details of a relationship have been worked out, the *Center* views it as a *de facto* functioning entity. "It exists, therefore it works." It can be a shock to them when they are confronted by a dissatisfied partner. They just forget sometimes that other people are not as emotionally self-sufficient as they are.

When *Centers* do turn their analytical gaze to their own relationships and family lives, they see much that can benefit from their organizational skills. Such precision cannot possibly be applied efficiently to every aspect of a relationship, but they will certainly try to make sure the basic daily routine is well worked out. Generally, though, if the *Center* in your life loves you, you will notice that life in your home runs like clockwork, with as little upset as possible, your children are aware of the difference between right and wrong and everything works or gets fixed quickly.

Centers will not rush into a relationship. The typical whirlwind romance is not for them. Their partners should expect a liaison that is slow to develop and full of analytical introspection. They will plan each move and investigate their feelings afterwards to see where they are heading. They will be steady, dependable and caring, when they can find time away from their career, which will be the major rival for any prospective partner.

Centers also have their innate talent for trail-blazing and focus, which will be factors in any relationship they form. They will succeed in relationships in the same way they approach the rest of life, applying logic and/or intuition as needed to solve personal problems without guilt or regret.

As the courtship develops, partners need to be aware of habit patterns that form. If they don't like them, this is the time to break them, before the *Center* makes them an integral part of the relationship.

Compatibilities

What are you looking for in a romantic relationship? Here are the major qualities to expect in a partnership when **Center** pairs up with one of the other BrainLines readings.

Center with Center
Deep awareness and empathy for each other's emotional needs. Rivalry over who calls the shots, until middle ground is reached. Profound connection, once the vision for the relationship is mutual.

Center with Balanced Double
Similar perceptions of the world.
Different methods of coping with problems.
Emotional volatility between the frenetic *Balanced*

Double and the introspective and determined *Center*.

Center with Clear
Hidden depths on both sides, the inwardly-focused *Center* and the low-emotion Clear.
Shared powers of concentration.
Different perceptions, the *Clear's* holistic view and the *Clear's* vision of personal success.
Superficial emotional contact.

Center with Right Clear or Right Double
Stability, as few surprises as possible.
Openness, if communication is made a priority.
Well-defined relationship with clear roles for each partner. Commitment to an orderly and long-lived union.

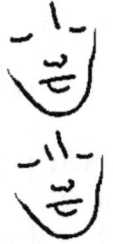

Center with Left Clear or Left Double
Romance, if the *Center* can appreciate it.
Unpredictability and originality, never a dull moment.
Control issues, *Left Clear* will want a voice, *Left Double* will balk at any limitations.
Active social life, the *Center* should be ready to party!

Work That Works

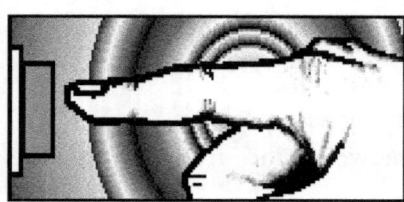

Centers often have the good fortune to have an early calling in life and are able to focus on that chosen field with enthusiasm. They have an extra intuitive dimension that combines a power of foresight with an extension of logical principles. They can therefore predict, using both hunches and educated guesswork. They have visionary power but, modestly, would not see it in that way. Just as both left and right brains are creative in their respective ways, so there is depth of perception on both sides.

Centers are focused, imaginative and innovative. They naturally seek outlets for their talents in areas where the possibilities seem most limitless.

Creative pursuits keep *Centers'* motors running. You will not find healthy *Centers* in careers that involve a lot of repetitious work, which would restrict the flow of input they need. Repetition of any kind that serves no purpose is unacceptable to *Centers*. Their distaste for superfluous activity is especially apparent in their preference for succinctness in speech and in the written word.

Centers are able to get to the core of a problem quickly, make pinpoint judgments and instant decisions, using their unique logical intuition. They have an eye for detail that translates to seeing relationships between what they are studying and what they already know. Taking those relationships and recognizing common features, they can apply new understanding to the problem with their melding of left and right brain strengths.

With their elevated level of consciousness and their balance of expression they are able to develop systems through which to mediate differences and overcome discord.

They tend to keep the lion's share of the work for themselves, only delegating tasks to people of whose abilities they are sure, keeping the final decisions for themselves. They typically have a precise idea of exactly what they want from their team and will get that information across in no uncertain terms.

They will stand behind their decisions until another, well-substantiated point of view, stated in calm and considered fashion, creates a new set of feelings that might sway their opinion, or until they come up with something themselves. When *Centers* learn something, they will want to share it, not have a long discussion of its veracity. They will allow emotional presentations to wash right over them without exposing themselves to input overload. Thus, the finer points of an opposing argument will be lost in the filtered-out noise, jeopardizing an agreement. *Centers* are equipped to use everything they perceive to create new ideas that may change their approach to any given problem. This flexibility can be important in a volatile marketplace (as well as life), where thinking on one's feet is sometimes better than deep research of each nuance and possible outcome of a decision.

Because of their naturally powerful presence and their self-sufficiency, they very often become fine mediators, using their interpersonal strengths and overview to enable them to perceive the conflicting desires and feelings of everyone involved. They can then formulate solutions and compromises that do not involve them emotionally. They may feel deeply about an issue, yet are able to keep their own external show of emotions in check, most of the time.

It is hard to hide such talent as theirs, so they tend to shine in the workplace, creating loyalty and admiration for their fairness and honesty. They prize honesty above truth. They may have different perceptions from someone else, but will respect another's honest opinion, even if they do not agree with it.

Although there are many *Center* politicians, due to their natural air of leadership and their verbal skills, they will not typically engage in spin-doctoring and hyperbole, if they remain true to their reading. If the public arena does not distort their natural preferences, they will avoid the ambiguous language and hidden agendas of political and diplomatic life. They admire simplicity and precision in everything, especially speech.

They might take their unique overview, originality and their concern for others to become enlightened and stimulating educators, bringing to their pupils the joy they feel for their subject. They would not seek out long-term connections with their students on a personal level, having little need for that kind of relationship. They will try to keep their students interested and entertained through their ability to access memory on a random basis. As they speak, the material they use will suggest connections to other subjects and their own experiences. When the *Center* teaches, no two classes will be the same.

No matter where *Centers* are found, they will be recognized for their calm, sensible approach to whatever involves them. As team members, they are a voice for reason, unemotional but not unfeeling, seeking the rational solutions, getting along with everyone and putting up with the emotional outbursts around them for the sake of the project.

While they are ambitious, they are not totally self-serving. The work that interests them will take precedence over their own advancement. They will naturally rise to their deserved place in the profession they choose, without riding roughshod over others, although their fellow workers may not see it that way at the time. Their straightforward style may seem arrogant to people who do not appreciate the honesty that drives it.

Centers are well-suited to the healing professions, especially those that are outside the Western medical mainstream. Holistic techniques will naturally appeal, because they fit their perception of the world as a single, intra-connected entity. No matter which field they are in, they will be caring and concerned, yet detached enough in their demeanor not to be overwhelmed. *Centers* may be found in any job in which measured and reasonable response is useful, from TV producer to surgeon, computer programmer to commercial airline pilot.

Along with their acute awareness of the world around them can come the vulnerability to over-stimulation of the senses, both physical and intuitive. They may find a noisy work environment not just disturbing but downright painful. The constant barrage of emotional and sensory input must be below their own tolerance limit, or they will be unable to operate. Their workplace may have to become a haven of peace for them from the outside world.

Strong-willed *Centers* are focused yet flexible free-thinkers, good at creating rapport in a team once trust is established, good at motivating and trouble-shooting, gifted with intuition and foresight, highly empathetic, emotionally controlled and usually self-confident. They are highly attuned to esthetic values and harmonious environments, both physical and mental. If they choose, they have great potential for leadership, using perceptiveness and creativity.

BrainLines of the Rich and Famous
Center

John Barrymore, Samuel Beckett, Jimmy Carter, Winston Churchill, Jean Cocteau, Lady Sarah Ferguson, M.F.K. Fisher, Lucien Freud, Richard Gere, Hal Hartley, Gregory Hines, Robyn Hitchcock, Anthony Hopkins, Augustus John, Michael Jordan, Stephen King, Jude Law, Ewan McGregor, Monty Miranda, Bishop Paul Moore, Ezra Pound, Tom Selleck, Amanda Seyfried, Igor Stravinsky, John Travolta, James van Praagh, Gore Vidal, Alan Wolk.

PERSONALITY PROFILE FOR *CLEAR*

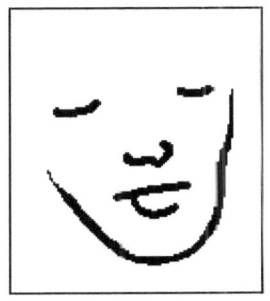

The Clear *reading can be recognized in people who, even when they frown, exhibit no clearly-defined pattern of vertical lines on the area between the eyebrows. They may show a jumble of partial lines, dimples and swirls, or no markings at all.*

Quick Summary for *Clear*

Read this to get a feel for what BrainLines can tell you about yourself and people you meet. The Full Profile begins on the following page.

You'll never feel the need for cosmetic surgery between your eyebrows. Maybe that's how you can be so focused on your goals. On the way, you're appreciating all the trappings of success for which you've worked so hard. You're designed to adapt to any circumstances, but, in adulthood, beware of becoming a hard-liner. You've left some egos floating in your wake, but it's a tough world out there and you'll think about the meaning of life once you've got where you're going.

Relationship tip: You're looking for the ideal mate in a stable relationship. The less emotional, the better, as far as you're concerned. Growing up, some dates may have called you insensitive, so you've learned to shape your responses to anyone's needs. The pressure of doing so will build. Better tell your partner how you feel before you explode. With your extensive abilities, you're a prize catch for any prospective partner.

FULL PROFILE FOR *CLEAR*

The Hard-Wired Brain

The *Clear* reading signifies someone who has equal access to both brain hemispheres and a low emotional level. Over time, a *Clear* may tend to be influenced more by one side than the other. Their ability to do this helps them blend with the crowd and learn the behavior patterns necessary to succeed in their milieu.

This fundamental adaptability is the great strength of the *Clears*, who do not have any dominant traits to sublimate. They can adopt the relevant traits of either hemisphere without emotional strain to become whoever they need to be. Clears are the ultimate survivors, self-centered and focused on success.

Clears are the backbone of society, the shapers of cultural norms, social mores and the political scene. They are a powerful sector and the focus that they can bring to bear on the goal of their own well-being and their position in society makes them committed workers and unrestrained consumers who fuel the economy of the country.

In short, most people with the *Clear* reading are good if somewhat guarded neighbors, stable family members and are diligent workers who enjoy the life and the luxuries that are their rewards.

Feeling the Feelings

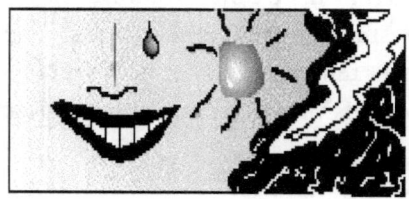

Clears in general act on a far less emotionally sensitive level than the other readings. Their reaction times are longer in new situations and they will not let sentimental issues cloud their judgment. They guard themselves from other people's emotional distractions in everyday life by observing the people they meet and judging their usefulness. Those who cannot advance the *Clears'* progress will be ignored or perhaps noted for future reference. In this way, *Clears* can operate efficiently, focusing on their own problems without exposing themselves to needless complications.

They are likely to have fairly conservative and traditional values and unless they feel the pay-off is big enough they will prefer not to go against the tide of public opinion, especially in their own community.

Clears typically do not allow their own emotions to govern their decisions. They have the ability to keep their feelings inside with no great strain on the psyche, and are wary of people who have to display their emotions openly.

They have a slow fuse when it comes to getting angry, but once that fuse has run out, it takes a long time to dispel the emotion they feel. They tend to hold grudges when they have been forced into anger or any other strong emotion. The combination of that tendency and their wish to avoid emotional confrontations can cause *Clears* to harbor ill-feelings when another person does something to cause them discomfort.

A *Clear* may try to avoid an upset by not making the other person stop doing whatever it is that is making the *Clear* uncomfortable. Emotions can then swell up until a bursting point is reached, when the *Clear* will explode, to the amazement of the unwitting target of their anger.

Clears are prone to such volcanic scenes, followed by a return to their calm demeanor, putting the quiet, unemotional damper on their feelings.

Social Ability

 Organization is the key for *Clears*, they like their lives to run according to plan, at a deliberate speed and course, with a certain goal in mind. They are wary of change, preferring to stay with the plan that they formulate or is formulated for them. *Clears* can be dogmatic about their beliefs when they feel they are being challenged, knowing that their views are backed by others who have done the original research and creative work.

Having a lower sensitivity level means that *Clears* are able to concentrate on the business of living and working without expending emotional energy on worrying about others. That is not to say they do not care about other people at all, but they may have to be reminded sometimes of the possible effects of their actions on other people's lives. Their focus on the goals they set is total, so they can give the impression of looking right through anyone that stands in their way.

Their social behavior will differ depending upon their left-brain or right-brain influences.

The left-brained *Clears* are conformist and retiring. They will only be openly friendly with a select few quiet people. They are not comfortable at social occasions or with small-talk. They have little desire to be placed in a leadership position due to their shyness. However, they will be loyal to those they choose as leaders.

Left-brained *Clears'* interests are fairly narrow and technically-oriented. They work best when they are given projects that pique those interests, but will also deal with unappealing work in a stoic manner. They are methodical and modest about themselves and do not force their opinions on others, often shielding themselves from other people's concerns. Left-brained *Clears* tend toward a pessimistic outlook on life, which stems from their need to examine every situation for its dangers and drawbacks.

For them, taking risks or venturing outside the norms to which they adhere is dangerous. These introverted *Clears* are the ones with the hidden emotions, which can explode without warning, release the pressure and subside again.

The more right-brained *Clears* are talkative, gregarious and popular. They are more status-conscious and tend to flaunt what they have. Their wide circle of friends are also likely to be *Clears*, who would appreciate and understand each other's motivations. They are more overtly emotional than the left-brained *Clears*, and more sensitive to other people's feelings, so are able to know what people want to hear and are more ready to assist them.

Right-brained *Clears* are more ready for change than the conservative left-brained *Clears*, more willing to talk about abstract ideas, more comfortable in an active social environment.

A dichotomy that *Clears* have to face is their need for approval from their community versus their aversion to putting other people's needs first. They may become involved in community activism at first to combat something that threatens their way of life and be drawn deeper into other issues with which they may have little sympathy, but which will bring them approval or put their skills to use.

A *Clear* may, for example, be involved with an environmental organization because of his expertise in a certain area, or because it seemed like the thing to do, and have little affinity for the cause. He might use his skills to organize an organic produce market, then go home and spray pesticide on his lawn, without seeing any irony in it. *Clears* simply tend to worry less than the other BrainLines readings about issues that they think do not directly affect their livelihood and safety. They prefer to trust the Establishment line on social and environmental issues, rather than expose themselves by speaking out.

Clears are attracted by organized religion with its ability to absorb the individual, the ease of its social interaction and its prescribed rules of conduct. *Clears* usually feel uncomfortable talking about metaphysical matters unless they have been taught well-defined methods of doing so. They are, however, comfortable with repetitive tasks, so the learning of rituals, doctrines, prayer and meditation techniques is not a problem for them. *Clears* will be found at every level of religious and spiritual organizations, thanks to their ability to immerse themselves in the learning process and their talent for teaching what they have learned and making it understandable. They will be devoted followers and workers on behalf of their chosen faith, which may induce them to become involved in community activities and charitable work.

Clears are generally solid citizens who seek to preserve the *status quo*, preferring business as usual to any disruptive exploration of new ways. They will be loyal followers of, for example, the political party that is predominant in their area, or of the religious persuasion of their parents. They are willing to pitch in and help, even volunteer, if it seems to be expected of them and if they will gain general approval for it.

Dating and Relating

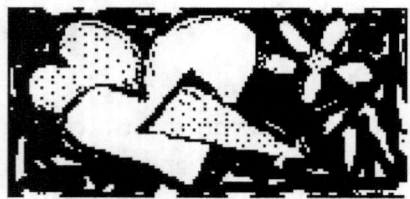

To commit to a long-term relationship, *Clears* are driven by their sense of responsibility and their needs for social conformity and stability. So, these are the people to choose for a relationship that holds little emotional turmoil, other than the occasional outburst.

When *Clears* have decided they want to settle down with someone, they will be true to their word and commit their time and resources to creating a marriage and a family. If they have happy memories of their early years, they will try to reproduce their childhood home as closely as possible. *Clears* make decisions based on prior experience, so, as with most situations in their lives, they prefer not to mess with a formula that has worked before. *Clears* are preservers of the classic picture of typical family life, striving to keep an even keel and working to make their home more comfortable and well-equipped.

Any of the other BrainLines readings would be compatible with a *Clear* partner, but only if they are willing to curb their impulse to rock the boat by trying to make a marriage conform to their own ideals. With some give and take on both sides, *Clears* and their partners with other readings can certainly influence one another. *Clears* will provide a secure environment in which the marriage can last, but will prefer not to have to deal with any emotional upheavals.

All the other readings would contribute their own strengths to a marriage with a *Clear*. The *Balanced Double* reading would contribute sensitivity and imagination; the *Left* readings would contribute compassion and intuition; the *Right* readings would bring their empathy and logical originality; the *Centers* would give the *Clears* room to maneuver, a respect for the spiritual side of a relationship and the benefit of their broad overview.

For some prospective partners, the compromises that must be made for the survival and growth of a marriage between *Clears* and the other readings may seem too much like giving up their identity, especially if there is any doubt as to their total commitment. The early stages of the relationship may be a struggle of wills, but, in the end, both parties should benefit. *Clears* will absorb some of the qualities of their spouse, while their spouses will gain the focus, stability and predictability of the *Clears*.

When *Clears* form a relationship with *Clears*, there is a match of personalities that will ensure a relatively smooth path for them both. Their marriage stands a good chance of being a long-lasting one, conforming to community norms as much as possible and without troublesome emotional turmoil. For extroverted *Clears*, this may be a little too mundane, but the survival of the relationship may take precedence over their search for excitement, which they may have to find in their careers or hobbies.

The interests of *Clears* are so widespread and their talents so diverse, that a relationship with a *Clear* can be stimulating for all readings. They can even learn to appear intuitive or emotionally sensitive by watching and understanding what their spouses want from them. They can learn to pick up signs of discontent in their partners and deal with the situation before it goes too far. This may be enough of a show of emotional sensitivity for those people with *Left* readings that are willing to compromise on the amount of sympathetic feedback they need.

Giving the positive reinforcement that the *Lefts'* need could be a part of the relationship that *Clears* would have to learn in order to satisfy their partners' requirements for loving support. A truly intuitive *Left* person would know that the *Clear* was simulating the proper responses, but might play along, appreciating how much effort their loved one was putting into bolstering their ego.

For the *Right* readings, who need less emotional support, a marriage to a *Clear* who can learn to recognize signs of distress and react sympathetically would add compatibility to a relationship in which both partners are fascinated by the details of life, with learning and using the information they gain in forming plans for the future.

Clears are good at being attentive to and showing patience with their partners and children, using their own experience of childhood as their guide. They will want to lay down rules as a set of parameters within which the family must operate, because that is the way they like their life to be structured. *Clears* like guidelines for the sake of stability, they are comforted by constrictions on their behavior, because they are uneasy with the unpredictability of human nature, including their own.

Their abilities will be sorely tried by a partner who is restless with the *Clear's* conformity or a child who wants to rebel against the *Clear's* rules. These kinds of situations will bring out the adaptive learning skills in *Clears*, who will show their love through their actions in trying to understand their spouses and children and discussing the best way of dealing with their problems. This will all take some time, as *Clears* are out of their emotional and intuitive depth in such circumstances.

If emotional demands are pushed far enough, a partner will witness an emotional outburst in which all bottled-up tensions will be released. Every past perceived slight will surface. The safest ploy is probably to stand back and watch the storm pass. Once the pressure is released, the *Clear* will be his old amenable self once more.

Their strength does not lie in making snap decisions about the future. They need to process the information they receive and compare it with their experience or what they have otherwise learned before they will be comfortable in decision-making. In times of emergency, however, they will be unemotional under pressure and be able to do the right thing, as they see it, quickly and as efficiently as they can.

Clears' plans will be strongly influenced by the word "should", as in, "I should do such and such a thing, although I don't really want to." Their sense of duty and need for stability will make them constant and loyal partners and parents.

Compatibilities

 What are you looking for in a romantic relationship? Here are the major qualities to expect in a partnership when a *Clear* pairs up with one of the other BrainLines readings.

 *Clear with Balanced Double.*
Sensitivity from *Balanced Double* to *Clear's* needs.
Lack of focus in *Balanced Double* may be challenging to *Clear.*
Compromise is necessary from both partners.
Working for a long-term relationship will be a priority.

 *Clear with Center.*
Spiritual and emotional stimulation for the *Clear.*
Room for both partners' interests.
Shared ability to focus, but different perceptions of success.

 *Clear with Clear.*
Smooth path, with few emotional drains.
Compatible focus on goals and relationship.
Possible control issues inside relationship.
Stable, long-term family life.

 *Clear with Right Clear or Right Double.*
Empathy for *Clear's* style.
Logical progression of relationship.
Similar focus on life goals.
Minor demand for intimacy.

Clear with Left Clear or Left Double.
Compassionate, emotional and sensitive *Left Clear* or *Left Double* brings new dimension to *Clear's* notion of a relationship.

Intuitive style of *Left Clear* or *Left Double* requires adaptability from *Clear.*

Compromise and deep commitment are necessary.

Work That Works

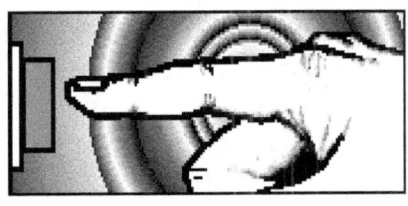

Clears are born survivors, thanks to their ability to focus on goals and rewards. The *kind* of work in which they are involved is not as important to them as financial security or the approval of their peers.

Clears tend to make decisions at a slower rate than the other readings. They like to study all sides of a situation, make a thorough assessment of the pros and cons and compare it with what they have already experienced. *Clears* will try to come up with what they perceive as a flawless plan or the perfect answer to a problem. They prefer to operate on experience and accumulated information, within prescribed guidelines and with proven precedents.

Aside from the kind of outburst that occurs after long-term brooding on a perceived slight, *Clears* will rarely be swayed by their emotions. They are too much the survivors, the calm-under-pressure assessors of risk and failure, to sacrifice their clear-headed, analytical abilities. Only when they are in a situation in which they must follow orders, such as professional sports or the military, will they act without thinking and suffer the consequences later.

Their preference for guidelines and prescribed modes of behavior makes *Clears* the ideal kind of person for disciplined environments in which they can strive for perfection. They are also suited for activities which provide an adrenaline high, perhaps as a compensation for their lower emotional level. Many successful team-sports athletes are *Clears*, as are many of the top racing drivers and experts in the 'thrill' sports, like surfing and hang-gliding, which demand self-discipline in certain areas of technique and equipment handling. *Clears* experience less fear through imagination of consequences than the more right-brained readings.

They will often surprise themselves by what they can accomplish, once the responsibility for analysis of a situation lies with their boss and they can allow themselves to act on impulse, but within the rules. In that kind of environment they can learn how to make quick decisions, using what they have been taught about the possible outcome of certain courses of action. *Clears* thrive in situations in which they can work 'by the book'.

There is, however, no occupation that a *Clear* cannot learn well enough to be at least adequately accomplished. They are found in all walks of life, learning how to be the best at surviving within the profession. Their focus is always upon success and reward for their efforts. They are not the self-effacing types who will put aside their own needs for someone else's comfort, unless persuaded to do so by social or religious pressures. Survival and security are so important to them that they will fight tooth and nail against anything that they see as a threat to the well-ordered systems in which they choose to operate.

Clears are deliberate thinkers when faced with new problems or change, which is why they prefer to have guidelines and precedents for dealing with situations, rather than having to act on hunches. When a *Clear* gives an opinion or solution it has been thought through and matched to their mental files relating to the subject. They will form solutions based on experience and on proven methods. While *Clears* have the adaptability to pick up techniques to help them cope with new situations, these are learned rather than intuitive methods.

Set routines are important to *Clears* and they like to start with precedents from which they can operate. They tend not to create truly original concepts, but work from a framework of knowledge to put things in order and make existing systems run better. *Clears* will learn how to do something well but will balk at creating new methods of doing it.

In general, *Clears* are ambitious and competitive workers and if they work in teams or on committees they take an active and vociferous part. They like to get along with their companions, and therefore look for the best way to work with others that will produce results. They like to be rewarded with praise and peer approval and will work hard to be seen as the best in their field.

These people have focus and are willing to learn a repetitive discipline very thoroughly, even a spiritual one, if they feel the rewards will be great enough. The rewards could be monetary or those gained through audience approval, power or fame.

Clears will want to control all aspects of a project, trying to get it done to their specifications, even if that means they take the whole workload on themselves. Being unwilling to relinquish control, they will not be good at delegating and may take on too much for their own, and the project's, good. This predisposition to work-aholism can be handled by designating the team with which the *Clear* must work and giving firm boundaries to their respective duties. Such parameters will make a certain area theirs alone. They will work to make it the best damned area it can be.

Clears should be given time to learn all they can about a subject so that they have the knowledge with which to create their plans. They are not intuitive risk-takers, but careful, analytical decision makers. They need time to think things out, weighing the pros and cons, so immediate action should not be expected from them in situations such as troubleshooting. Given time to prepare, they will make the best possible decision and give the most helpful advice they can devise.

Yet all the points in the previous paragraph can be changed by the learning potential of the typical *Clears*. They are able to learn techniques thoroughly, even those that seem more suited to other BrainLines readings. Given the time and the proper instruction, they can learn to read people's emotions, make decisions based on informed hunches, adopt a caring bedside manner, give advice on spiritual matters, anything for which techniques have been created that can be taught.

Thus, there are no occupations to which *Clears* cannot mold themselves. They are involved in every facet of society, learning, accomplishing, succeeding, looking for security for themselves and their families, doing whatever it takes to get ahead.

BrainLines of the Rich and Famous
Clear

Celia Alvarez, Tammy Faye Bakker, Pierce Brosnan, William F. Buckley Jr., Chiang Kai-Shek, Lawrence Durrell, Antoine Fuqua, David Geffen, Benny Goodman, Ernest Hemingway, Djimon Hounson, Jesse Jackson, Rinko Kikuchi, Larry King, Henry Kissinger, Norman Mailer, Henry Miller, Benjamin Netanyahu, Norio Ohga, Seth Rogen, Thierry Roussel, Mark Ruffalo, Bob Vila, Kurt Vonnegut, Barbara Walters.

PERSONALITY PROFILE FOR LEFT CLEAR

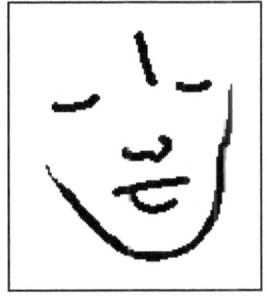

The Left Clear *displays one line on the left-hand side of the forehead between the eyebrows.*

Quick Summary for *Left Clear*

Read this to get a feel for what BrainLines can tell you about yourself and people you meet. The Full Profile begins on the following page.

You're a poet with a surgeon's eye. You are most successful in imaginative, even idealistic, pursuits. If you're an artist, you're the rare kind that has the confidence and pragmatism to market their own work. If you're an entrepreneur, you want to make people's lives better with your inventions, even while making a living from them. Don't hide your soft side, there's room at the top for a big heart.

Relationship tip: "I'm sorry, it just feels like we're getting in a rut." Did you say that? Love is your bed of oysters and you want to find the pearls. You've been disappointed when your vision of a relationship didn't come true. Once you've found someone who can keep up with your need for excitement, you'll devote yourself to refining the relationship until it's running smoothly. Then you'll be able to work towards your dream. Half the fun is getting there.

FULL PROFILE FOR *LEFT CLEAR*

The Hard-Wired Brain

The people with the *Left Clear* reading have creative originality and idealistic interests combined with pragmatism and focus. They will show some of the same tenacity and relentlessness that the *Clears* display when it comes to survival. Survival to those with either a partial or a pure *Clear* reading may mean simply maintaining or increasing the comfort level to which they have become accustomed.

They are highly aware of the subtleties of their environment and are attuned to abstract subjects, such as spirituality, esthetics and the emotional health of themselves and the people around them. The *Left Clears* are constantly bombarded by input on these subjects. They can accept this incoming information as an emotional stimulus, then process the thoughts that arise from it.

The *Clear* influence gives some pragmatism to the input, guiding it into creative and original concepts that will be useful to them in their careers and personal lives. In this way the two parts of their reading complement each other to create a personality that is sensitive to issues concerning the more abstract side of life and that uses those issues in a practical way. Everything a *Left Clear* learns is added to the store of personal experience they use in making everyday decisions. *Left Clears* rely on their memories of past experience and the feelings that are created by these memories to mold their opinions and decisions.

Left Clears are alert, sensitive, ingenious, with highly original ideas and the drive to see projects through with focus, if they can really feel personally involved. If faced with a repetitive or routine task, they will have to be convinced that it is truly necessary to submit themselves to the boredom. Once they are persuaded that the favorable end result depends on the repetitive task they will grit their teeth and grind through it. To accomplish this, the *Left Clear* will have to be given one thing to do, with a clear goal and no distractions from other people. When faced with a tedious task, they will look around for something to divert their attention and people are their favorite diversions.

They are good at making snap decisions in risky endeavors, or when they have many choices that are equally sound. That is when their intuitive skill, with which they are completely comfortable, kicks in. The *Left Clears* are able to access their instinctive 'feel' for situations and people, seeing the choice they make as part of the greater project. While there may be no apparent logical reason behind these decisions, they are usually just as valid.

Feeling the Feelings

Left Clears have deep feelings, high emotions and spontaneous decision-making. If they encounter resistance to their decisions, they will be willing to modify them. A word of advice, though - if you are the one trying to change their course of action - be prepared to take the blame if things go wrong. They are always thinking on their feet, going with the course of action that occurs to them on the spur of the moment.

Left Clears have a reputation for being mercurial in their emotional states, mainly because they can be so focused on something that any outburst is a shock. They tend not to hold anything back and can blow hot and cold, playing to their love of the dramatic in life. They are spontaneous and imaginative, colorful and humorous, with a gift of persuasiveness that comes largely from their empathy for others' emotions and the passion they convey.

They are sensitive to others' moods and dislike conflict to such an extent that they will do something the way someone else wants it done, for the sake of harmony, even though they may disagree with the method. If the goals remain the same, they may go back to doing it their own way and present their solution with a somewhat defensive air. If they are constantly forced to do things the way others want them done, at the expense of their own creative skills, they will bottle up the resentment until the inevitable emotional outburst occurs.

Social Ability

 *Left Clears* are peacemakers who prefer to see the similarities between people rather than their differences. They will try to find the common ground between two sides and effect a compromise. Their ability to read between the lines and the sensitivity they have for how others are feeling help in their mediation efforts, as do the *Clear* qualities of considering the facts to reach logical conclusions, although this takes more time.

Multiple interests, a sense of humor and a wide perspective make *Left Clears* interesting and stimulating company. They are warm, funny and friendly, giving freely of their emotions and personal feelings, which can also mean that being around them may be a little *too* exciting for some people, when a squall of emotion passes through.

Their all-inclusive view of life and their iconoclastic bent may manifest themselves in their arguing just for the fun of it. The mental stimulation in discussion, their interest in other people's emotional involvement in various subjects and their penchant for deflating egos and bringing humor to an overly serious discussion may make them take a position that is counter to their true beliefs. This does not show duplicity, however. *Left Clears* are secure enough in their overview of life and tolerant of other people's views that they can play all sides just for fun, while maintaining their beliefs and defending them when the situation gets serious.

Left Clears are likely to be interested in projects that they see as beneficial to the community or the world. This stems from their concern for other people's welfare. They will assume leadership easily and effectively in such causes. In most cases, although the *Left Clear's* greatest enjoyment comes from being around people, they will prefer to be in charge, running things their way, so that there is less chance of a project they care about getting out of their control. If they work with a team, they will surround themselves with people who will feel as strongly as they do about the project.

Whatever is important to them will become a driving concern to which they will direct their focus. Combined with their original and creative mind, this focus will make them a formidable force in any field to which they turn their attention.

Left Clear people are intuitive and idealistic seekers of the greater truths in life, with a penchant for spiritual matters. Their ability to see life as a whole, rather than a succession of separate events, gives them a valuable universal overview. This long-range vision, coupled with their focus, helps them to anticipate events and set goals.

Dating and Relating

Left Clears are emotional and compassionate people, on the whole. Relationships of all kinds are of primary importance to them, perhaps partly as a validation for their non-conformist and original ways. They need to hear from people around them that they are loved and understood. This need may stem from their uneasiness with behavior that is outside the norm, which is exactly the kind of behavior that is encouraged by their creative urges.

They will want to create a romantic aura which will emphasize intimacy and deep bonding, while their cool, calculating other side keeps an eye on the strategy's success. If there is the probability of emotional pain, they will call a halt to the proceedings. One by-product of this is that it lessens the distress that occurs when their partner does not seem as interested in the relationship as they are.

Left Clears will initiate conversation in order to connect with their partners. They will consider their partner their best friend and expect empathy over a concern or upset. Although they are emotional, they will try to rationally analyze conflict with their partner. They are able to tolerate diverse emotions and understand that everything does not need to make perfect sense. This is done by using questions as tools, in empathetic, emotional language, to determine how problems may have arisen.

To keep romance alive, the *Left Clear* will devise creative ways such as flowers, cards, poems or playing with shared objects of intimacy to kindle and stimulate the liveliness of their partnership. Even with their romantic and compassionate nature they are practical enough to realize that love is not enough. They know that a committed relationship requires many skills which have little to do with their emotional bond.

While they may expect their partner to share their intuitive abilities, their pragmatism will help to dispel the illusion that lovers should always know what the other needs, without asking.

If *Left Clears* were prepared to meet the challenge of their opposites and were willing to sacrifice quality intimacy time in order to gain a balanced creative collaboration, they could be compatible with *Right Clears*. The time they spent working together on their collaborative achievements would have to take the place of simple domestic togetherness.

An easier path and greater chance of success might be found with someone with the same reading, a *Left Clear*, a relationship in which they would both be comfortable with each other's non-conformity and which would bring them greater intimacy. However, they would both have to give each other a great deal of approval in order to balance out their sensitivity to criticism.

Other alternatives would be a *Left Double, Balanced Double* or even a *Center* for greater understanding of their emotional needs and imagination.

Left Clears are able to handle many details of a relationship and home life at once. They will prepare a dinner party, talk on the phone and organize their calendar for the week while changing diapers. They may make lists of their tasks, but will attend to them in a seemingly random way, as the mood strikes them, rather than devising a rigid order of completion. *Left Clears* rely on synchronicity more than they might like to admit. They will complete the tasks in their own fashion and congratulate themselves on their organization.

Left Clears find schedules, deadlines and routines hard to deal with. They are more comfortable with unpredictable quirks of fate and working on hunches, whatever feels right at the moment. They will try to make it to a dinner date on time, but they are always willing to be side-tracked by some interesting shift of plans.

A relationship with a *Left Clear* will be exciting in its spontaneity, its emotional level and its air of confusion. The furnishings in their home will be chosen with an eye to esthetics, rather than pure function and, during courtship, are likely to change according their whim and how they want their 'stage' to be set.

Left Clears rely on what their feelings tell them about people, places and things. If they have a sense of unease, their survival mechanism will take over. They will pick up their partner's moods and needs without even thinking and will want to talk about them. A *Left Clear's* partner should be prepared to have their feelings laid bare and examined regularly.

In conclusion, *Left Clears* will be romantic, vulnerable, sensitive, intuitive partners, willing to find the flaws in a relationship and fix them, but unwilling to be bored. Do not expect them to stay in a partnership that does not excite them. Emotional stimulation is the key to a successful relationship with a *Left Clear*.

Compatibilities

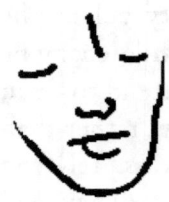

What are you looking for in a romantic relationship? Here are the major qualities to expect in a partnership when a **Left Clear** pairs up with one of the other BrainLines readings.

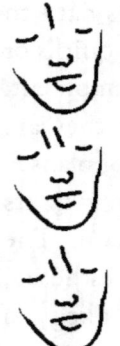

Left Clear with Center, Balanced Double or Left Double
Emotional understanding.
Imaginative pursuits and similar interests.
Unpredictability.
Excitement is the key to this relationship's longevity.

Left Clear with Clear
Reality and pragmatic influence.
Stability and commitment are priorities.
Careers and outside interests are important.
Emotions are somewhat stifled.

Left Clear with Right Clear or Right Double
Challenge in relating to partner's style, especially in criticism.
Rounded creativity creates diverse interests.
Routine becomes part of the relationship.
Compromise necessary on both sides to avoid conflict.

Left Clear with Left Clear
Intimacy and complete understanding of style.
Harmony comes from similar types of interest.
Romance is important.
Spontaneous and unpredictable relationship.

Work That Works

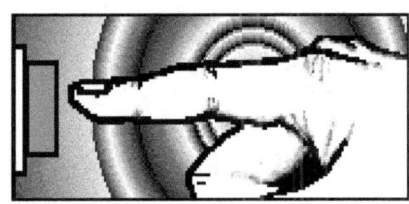

Left Clears are best at working with a group of people with whom they can exchange ideas and get feedback on the decisions they make. They are the pre-eminent networkers who will employ the expertise of other people to make up for their weaknesses in certain areas. They are generally conscientious and try to avoid mistakes and forgetfulness, so delegating tasks is one way for them to keep their minds clear.

Their main strength is in making quick, intuitive decisions, relying on their feelings and their reading of other people's emotions. *Left Clears* will not ask for too many details up front. They make good project managers and troubleshooters, using their 'emotional radar' to pick up on possible causes of friction between people and smoothing any ruffled feathers before the problem becomes unmanageable. They are always looking for harmony in everything that affects them, both emotionally and physically.

Left Clears make snap judgments, but they are not stubborn about sticking to those decisions, especially when they have thought them over for a while and allowed their *Clear* side to consider any drawbacks. These are flexible people, who will listen to input from others and be ready to modify their procedures, if they feel the change to be a positive one. This flexibility may cause them to get a reputation for being wishy-washy, although they will dig in their heels if they feel that the criticism of their methods is unsound.

Usually, *Left Clears* have a sixth sense about which way to go when deciding on a course of action. But snap judgments can go awry, especially when they arise from someone else's suggestions and there is no time to consider the drawbacks. Again, be prepared to take the blame if you have changed their mind and something goes irretrievably wrong. The *Clear* side's survival instinct will kick in and the *Left Clear* will not want to stand alone.

When they become interested in a project they will devote themselves to it, with the focus of the *Clear* side. But they may not be able to keep from tinkering with it, becoming absorbed in looking for new ways to make it 'perfect'. For that reason, they often need direction and deadlines to keep them focused on a goal, rather than on the process itself. However, the way in which direction is given can be a very touchy subject. *Left Clears* will tend to balk at direct orders and advice given in a tone that they find officious.

Tone of voice, body language, manner and phraseology are all meaningful input to the *Left Clears*, forming an emotional context for their response. Directions given to them need to be carefully considered and modulated, lest a conflict should arise from an innocent comment that was thoughtlessly expressed. They have the tendency to take any perceived criticism personally. It is not that they are thin-skinned, they just cannot help how their feelings are shaped by their environment. Encouragement, positive reinforcement, praise for the work already done on a project followed by suggestions for the next phase will be much more productive than a direct order from on high.

Why bother with what some people may see as needless molly-coddling? Because, when their strengths are used and encouraged, *Left Clears* are extremely effective at dealing with people, using their enthusiasm and empathy to form bonds with customers and workmates. They work well at brainstorming ideas with the group, gathering suggestions and acting on the feelings they get from the input. Their innovative minds can create ground-breaking ideas and connections that others might never consider, and may make the difference between success and failure.

They may be prone to selective listening by filtering out what they consider extraneous information, hearing only what they want to hear. Their focus on their own success may lead them to withhold information (since information is power), which may erode the *Left Clear's* self-esteem, if they feel they cannot freely express themselves.

Their ability to focus on a chosen path will counteract their seeming lack of commitment to one idea. It must be remembered that a *Left Clear's* focus is broader than other people's. They will commit to the job as a whole, while always looking for an exciting new way of accomplishing it. To others, this will seem like lack of concentration. However, to a *Left Clear,* the end is as important as, if sometimes less interesting than, the means.

Their methods may seem alarming in what has traditionally been the linear environment of the business world, especially their learning style. They tend to pick up information on a subject as they are explaining it. Armed with a few basic facts, they will start to visualize the information, seeing nuances and connections in the process. They will want to share these discoveries with someone else and, as they put their images into words, new images will spring to mind. This can happen at any stage of a project. A random thought will suggest a new line of inquiry or a different approach to a problem, which itself may lead to more discoveries. The well-researched, considered and logical approach is not for *Left Clears*. Their great strength is their ability to change pace and direction while aiming for the same goal.

The education profession is very suitable for *Left Clears*, who are best at occupations that use their interpersonal skills, their abstract creativity in decision-making, their visual or musical propensities or their penchant for spiritual investigation. A class with a *Left Clear* as a teacher is liable to contain more illuminating story-telling than a recitation of facts. They are likely to be attracted to art, ecology and religious, spiritual or personal counseling.

If *Left Clears* should choose the medical profession, they would find themselves most comfortable in family practice, pediatrics, geriatrics or other specialties in which a long-term professional relationship can be created.

Their interpersonal contacts are so important to them that their relationship with their clients may mean more to them than the service they are providing, but that might give their business the edge over the competition. These people form friendships, not just business relationships, which can go a long way in smoothing out any differences on project teams, in human resource departments, or in customer relations.

These are innovative, spontaneous, sensitive people, whose skills with people and delegation can be an asset to any organization willing to recognize the value of the *Left Clears'* approach.

BrainLines of the Rich and Famous
Left Clear

W.H.Auden, Lindsey Buckingham, Bill Cosby, Ram Dass, Willem de Kooning, Robert J. Eaton, Gerald Ford, Bill Gates, John Goodman, Scott Hamilton, Peter Jennings, Minnesota Fats, Rupert

Murdoch, Sarah Polley, Tim Robbins, Chris Rock, Richard Serra, Isaac Bashevis Singer, Gertrude Stein.

PERSONALITY PROFILE FOR LEFT DOUBLE

The Left Double *displays two lines, one on each side of the forehead between the eyebrows. The line on the left-hand side is longer than the one on the right-hand side.*

Quick Summary for *Left Double*

Read this to get a feel for what BrainLines can tell you about yourself and people you meet. The Full Profile begins on the following page.

"There are no pastimes, only passions." Bet you wish you'd said that. If you've been given a free enough rein in your lifetime, you feel the truth of that saying. There's a neon sign over your head that reads, "Creative." And that can cover everything from painting through writing to social skills and lifestyle. You are aware of all the emotional levels around you. If you could, you'd ask everyone the questions that would lay their souls bare. You are spontaneous, sensitive and emotional. You're quite a handful!

Relationship tip: Courtship, to you, means mystery and theater. Not to mention passion, which is a constant in your life. Your partner will have to be imaginative to keep up with you. You also need someone who is not intimidated by non-conformity and creativity, and who is not easily embarrassed. You may have to "upgrade" partners a few times, but the search will be exciting, delicious and full of fun. Just like life itself.

FULL PROFILE FOR *LEFT DOUBLE*

The Hard-Wired Brain

 People with the *Left Double* reading can be characterized as having many passionate interests, one of the favorites being social interaction. They are emotional, spontaneous, and imaginative. They feel constant input from all around them and have trouble ignoring any of it. This constant stimulus can cause them to want to act on everything they feel, sometimes without thinking their actions through. In the *Left Double's* case, this may be the best way to act, because of their deep intuitive ability. Self-actualized *Left Doubles* will have learned to pay attention to their inner voices and use them trustingly.

All their thought processes start out with their creative, following which, if they allow themselves the time, their linear will contribute its logical side to the solving of problems. If truly objective decisions and answers are required, *Left Doubles* need to be given the time to consider the questions so that they can devote their entire thought process to them. Their primary strength, however, is in intuitive decision-making. This may be a scary concept in the primarily linear business world, but a *Left Double* who is experienced in the field and is comfortable with intuitive power will be remarkably accurate in feeling out situations, especially those directly involving other people.

Left Doubles will get down to working on something with speed and determination. They are capable of organizing their resources quite efficiently and in a systematic way, at the same time bringing idealism, insight, persuasiveness, and a sense of the dramatic to bear on the issue at hand. This can be a formidable approach to any situation, overwhelming to the more staid readings, but an exciting roller-coaster ride for those who can hang on.

Left Doubles form deeply-felt opinions. Given time, they will perhaps reconsider their actions and other points of view. While they can be flexible about their actions, willing to change course and destination depending on the conditions, they are more likely to stick with their opinions, which, though quickly formed, become deeply ingrained.

Left Doubles think in images rather than words, so, while they may be able to hear and appreciate the views of others, their own views are so much more powerful and personal to them, because of the imagery they have constructed from their ideas and experience, that they are reluctant to change them.

People with this reading are often described as lacking in concentration, because they can divert their attention to many things, one after the other. They will have complete fascination in one endeavor, but if they are reminded of one of their other interests, they are able to shift their enthusiasm immediately. *Left Doubles* feel so intensely about everything they do, that people with other readings may not understand how such passion can be divided and still leave enough for them. That seemingly limitless enthusiasm is both a blessing and a curse for the *Left Doubles*. Without it, they would be hopelessly overwhelmed by the stimuli they receive from all around them, yet such intensity cannot be maintained without risking burnout.

They are intolerant of old ways of thinking if they feel these are non-functional and not progressive. *Left Doubles* are highly creative in finding new approaches to old problems, often seeing associations between concepts that others have missed. Their inventiveness can bring them to conclusions that are far from the accepted norms. It stimulates far-reaching consideration of subjects that may influence other decisions and lead the *Left Doubles* into another interesting situation. This is the pattern for *Left Doubles*, who can use their vision of the 'big picture' to provide endless outlets for their attention.

No matter which field *Left Doubles* choose, they will push for progress and enlightenment, using innovative techniques, persuasion and vision in the cause.

Feeling the Feelings

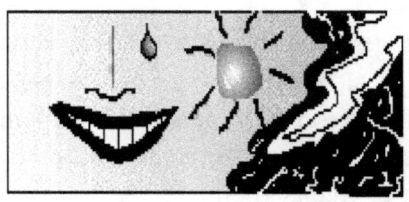

Making a *Left Double* stand back and recharge their batteries for a while can be difficult, since they tend to get so absorbed in their interests that their own welfare seems of secondary importance to them. This absorption may also give them the reputation of being unsociable at times, yet it is the passion that they have for what is important to them that gives these people the capability of getting involved in so many interesting and creative projects.

One of the things that *Left Doubles* find most difficult to take is criticism, or what they perceive as criticism. Everything they do is so based on emotion and feelings that they look for constant affirmation of their efforts. Criticism can seem to them a personal attack, which can cause a passionate outburst or foster resentment if they try to hold their hurt inside. They will often let the resentment build until they feel unappreciated by those around them, perhaps causing a loss of confidence in their own abilities. This can cause a shutdown in the *Left Double*, a state in which they are unable to make decisions for fear of disapproval and failure.

In the search for approval, which *Left Doubles* need on an emotional level, they will often make commitments that they are unable to fulfill. In the predominantly linear business world, this can lead to a reputation for not following through, or even lacking integrity. Their intention, however, is always to make people as comfortable as they can, even if that means making impossible promises.

Left Doubles usually have a well-developed sense of humor which they need to counteract their propensity for worrying. They will worry over little things as well as the greater problems besetting the world, which gives them a leaning towards getting involved in political and social issues. They tend to take these on as a personal challenge, which can make them emotionally vulnerable to criticism. *Left Doubles* have such clear images of what needs to be accomplished that they may be impatient with people who will not see their side of the issue. At times like that, *Left Doubles* need their sense of humor so that they can step back and see themselves getting too emotionally involved for the good of the discussion.

Their iconoclastic streak may make them rebellious and outspoken. They are defiant of rules that they deem unimportant but they will stop short of crossing any lines that will make them feel dishonest or unethical, according to their own credo. *Left Doubles* will be ethical and honest in business because they do not want to inflict pain on others, nor be burdened by guilt. Such stimuli would tax their sensibilities too much.

Social Ability

Left Doubles share the sensitivity to others with all the readings that display BrainLines, but to a greater degree. They are very responsive to other people's emotional states and try to act with the feelings of others in mind. This makes them vulnerable to the world around them, because they open themselves to others as they come to understand and empathize with them.

They find it difficult to turn off their "sympathy switch", always wanting to give advice and running the danger of being overloaded with other people's problems, but that does not deter them from being sociable, because they love the input. If *Left*

Doubles' ever do seem unsociable, the most likely reason is that their schedules are so full of work and personal appointments that they simply cannot find enough time.

Left Doubles tend to act on their idealistic impulses, which often leads them into volunteer work for the community or for what they see as the common good. They can be obstinate in defending their ideals and not backing away from conflict. Their natural rapport with other people, and their flexibility and charisma can inspire others to great efforts. *Left Doubles* are adept at leading groups and discussions, bringing their left-brain talents to bear in their verbal and organizational skills, while their sensitivity helps them bring their passion to a project without steam-rolling everybody else.

If *Left Doubles* say they are not interested in something, they mean it. They have no patience or time for anything that does not stir their soul. They find repetitive work intolerable and, if they feel they are involved in something that feels physically or psychologically detrimental, they will become so disheartened that they will be unable to carry on. At times, their reactions may seem overly dramatic, but they can tell when their energy is being sapped and wasted, and are not afraid to say so. They know that there are so many things that interest them that it seems wasteful to get involved in something that causes them distress if they do not believe whole-heartedly in it.

Left Doubles are often deeply concerned with the meaning of life and how to incorporate their conclusions into their everyday existence. They have the need for spiritual understanding and may be drawn to organized religion if they feel it satisfies their thirst for knowledge. If they were drawn to it, the doctrine would have to allow them plenty of freedom to think for themselves. The *Left Doubles'* grasp of abstract concepts makes the spiritual search appealing to them, especially if imagery is involved in the process rather than theoretical treatises that must be memorized.

Atmosphere, ambiance, esthetics, environment, *feng shui*; these are all important words in the *Left Double's* vocabulary, and they will often be found in volunteer organizations dealing with Nature issues. When *Left Doubles* have the appropriate backdrop, they can concentrate with less distraction on the images they use to figure problems out.

These are friendly, flexible, creative, original-thinking people, sensitive to all that goes on around them, while striving to become architects of change.

Dating and Relating

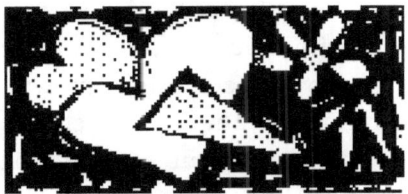 In order to fulfill their creative and emotional needs and to be understood, *Left Doubles* must have partners who are also very imaginative. Unless they have support and understanding for their non-conforming thoughts and actions, they will feel as if they are not being taken seriously. Since they have probably already had experience of being thought of as eccentric earlier in life, it can be a problem when they feel belittled or ignored. Conversely, their innate non-conformity can be a problem, even an embarrassment, for those partners who need more stability and are more wary of risk.

Their best choice for emotional compatibility would be someone with similar BrainLines, another *Left Double*. In such a relationship, there would be more likelihood of forming what is, for *Left Doubles*, the most desirable ingredient of a long-term relationship: a creative collaboration that sparks spontaneous fulfillment for both.

If *Left Doubles* have found enough support in the past for their left emotional side that it does not need reinforcement from their partners, their practical side might influence them towards a more stable and conventional relationship. In that case their choices would be a *Left Clear* or a *Balanced Double*. A *Center* would be grounded and focused enough to enjoy their 'spark' and not be intimidated by it.

If they can accept that they may never find a peaceful balance in their lives due to their divergent and expanding interests, they might realize that they need more than one long-term partner, in order to fulfill their ever-changing needs. *Left Doubles* can see the future possibilities of many courses of action and are always looking for ways to make their relationships better, which may sometimes mean that they have to look elsewhere for improvement. That constant search for better, more satisfying modes of relationship can be as challenging for the *Left Double* as it can be for the partner.

The partner in a relationship with a *Left Double* will have to show some commitment to improving emotional growth, learning new interpersonal skills, responding positively to excitement and change and participation in a working collaboration. *Left Doubles* can take criticism personally and may feel that they are under attack for their conduct in a relationship. They require deep understanding, tact and involvement from their partners, who will need to be prepared for emotional outbursts.

Left Doubles bring intensity and commitment to their love affairs, whether short-lived or lifelong. That can be a challenge for less emotionally volatile partners. While others are taking a 'breather', relaxing and allowing complacency and satisfaction to take hold, the *Left Doubles* are planning tomorrow's outing, menu, phone calls, as well as thinking about next summer's trip and wondering why their spouse is sitting over there, reading the paper and not showing affection or interest.

Left Doubles will try many things to keep romance simmering: gifts, flowers, poems and improvised events that will stimulate the energy of the partnership and keep it interesting. However, they have enough left-brain influence to know that love does not conquer all difficulties and that a committed relationship needs practical skills and a sense of responsibility and caring.

Left Doubles also need to be aware of their own romantic and trusting nature that can make them vulnerable to betrayal, from which they are usually saved by their intuition, if they pay enough attention to it. They are sensitive and open, willing to disclose their feelings completely. When problems arise, *Left Doubles* have original responses and advice to give their family members, giving freely of their overall vision and their practical side at the same time.

Positive reinforcement, a view of the future of the relationship and a willingness to relinquish center stage are all required of the partner of a *Left Double*. If you are entering such a partnership, here are some hints: be prepared for fireworks that will thrill and scare you; expect the unexpected; be ready to change plans at a moment's notice; show your concern by voicing your doubts, but try not to make them sound like criticism.

The partner's rewards for these efforts are creative collaboration, an exciting lack of routine, lots of humor and a commitment to a stimulating, evolving partnership.

Compatibilities

What are you looking for in a romantic relationship? Here are the major qualities to expect in a partnership when a *Left Double* pairs up with one of the other BrainLines readings.

Left Double with Center
Freedom of expression for both partners, anything goes. Control issues may arise inside the relationship. Spiritual growth may be a mutual priority.

Left Double with Balanced Double or Right Clear
Stability in the relationship.
Understanding of views and actions.
Emotional volatility.
Deep bonding.

Left Double with Clear or Right Clear
Little emotion may mean boredom.
Creative challenge for *Left Double*.
Different methods of problem-solving.
Compromise will be necessary.

Left Double with Right Double
Understanding of thought processes.
Practicality of *Right Double* strengthens relationship.
Emotional bond through similar traits in different concentrations.

Left Double with Left Double
Creativity on all levels.
Spontaneity makes for an exciting courtship, challenge later., Sensuality and romance are very important.
Long-term stability once the partners are deeply involved.

Work That Works

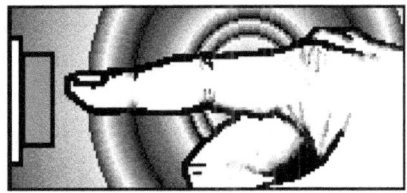

Left Doubles who have confidence in their abilities can make quick, intuitive decisions, acting on hunches and feelings. When they allow themselves to trust their intuition they will more often than not make the correct decision, thanks to their great sensitivity to input on all levels. With their brain in 'receive mode', they pick up highly diverse information, which they deal with mostly on an emotional level.

Their first instinct is to go with their hunches but, should doubts arise, *Left Doubles* have the capacity for re-examination and consideration of their snap decisions, using their analytical abilities to examine the details, possibly causing a change of approach. This can lead to the appearance of waffling, because they are so open to different ways of reaching their goals, but *Left Doubles* will stick to their emotionally-derived, deeply-held opinions. Just don't expect them to focus on a single procedure.

If there are strict boundaries of method and conduct, the *Left Doubles* will feel too confined to give the work their best efforts. They must be allowed a certain amount of freedom to operate, so that the workplace can benefit from their originality, tempered by their aptitude for logical progression. They have the ability to perceive the project as a whole, and, given time, to use their creativity in analysis and long-term targeting.

Because their feelings rule the way they react to the world, *Left Doubles* perform most enthusiastically on projects that are proposed to them in a positive manner. Continuing positive reinforcement is the key to maintaining their interest, with suggestions being given, rather than direct orders. Their pragmatism helps *Left Doubles* to see the logic in a chain of command, but they are still apt to take criticism somewhat personally.

Left Doubles are most comfortable in careers that give them close involvement with others. They love to be around people, finding their greatest interest in what makes people function the way they do, which may lead them to ask questions that others would be unwilling to ask. People to whom they are talking may feel that they are the subject of research. The more a *Left Double* knows about someone, the happier they are. This can mean that they are the people to go to when you want to know the latest office gossip. Gossip is information on the process of change, or interpersonal information, as far as *Left Doubles* are concerned.

They are always aware of other people's emotional states and can sense feelings through body language and other subliminal clues. They are, therefore, useful in team situations, which can benefit from the *Left Doubles'* sensitivity in making discussions and working arrangements run smoothly. They can tell when someone is uncomfortable, and will not feel right until everyone is as close as possible to feeling at ease.

Their preference for working with others can help to take the load of trying to balance their mental activities. They are adept at using team-members' strengths to work towards the set goal of a project.

Their amiability and powers of persuasion make others feel comfortable with giving their all, and *Left Doubles* will exhibit an infectious enthusiasm for a project that holds their interest. Their desire for harmonious surroundings make *Left Doubles* suited to areas that have to do with personal counseling, environmental protection, troubleshooting organizational problems and human resources issues.

The visual skills that *Left Doubles* possess will often lead into artistic fields which allow them freedom of expression without using a technique that would seem too limiting. Thus, photography might not appeal to them as much as painting. For painters, abstract imagery will appeal more than realism. Their creative component would enable them to deal with the technical aspects, but they might feel constricted by them. Whenever boundaries exist to curtail their expressive nature, *Left Doubles* will try to find a way around or through them.

BrainLines of the Rich and Famous
Left Double

Richard Burton, John Cleese, Glenn Close, Jean-Michel Cousteau, Judi Dench, Bob Dole, Lawrence Ferlinghetti, K.D. Lang, Andie McDowell, Marilyn Monroe, Jack Nicholson, Gregory Peck, Pablo Picasso, Paul Reffell, Chris Rock, Amanda Seyfried, Donna Sheehan, Kiki Smith, Randy Travis, Sol Wachtler, Boris Yeltsin.

PERSONALITY PROFILE FOR *RIGHT CLEAR*

The Right Clear *displays a single BrainLine on the right-hand side of the forehead between the eyebrows.*

Quick Summary for *Right Clear*

Read this to get a feel for what BrainLines can tell you about yourself and people you meet. The Full Profile begins on the following page.

The speed of your mind sometimes leaves others in the dust. You make quick decisions based on facts, and you don't let emotion slow you down. That cool exterior doesn't fool your closest friends and family, who know the love and compassion of which you are capable. So what if you've never been the life of the party, the world needs your laser-like ability to cut through the hype. You are a lighthouse in a stormy sea: dependable, safe, solid and true.

Relationship tip: Remember, it's O.K. (but not mandatory) to lighten up with the people who love you. You can show love in many ways other than programming the VCR or making picture-perfect pies. The right relationship must be as important as your career. Make allowances for the unexpected while drawing up the plans for your life together.

FULL PROFILE FOR *RIGHT CLEAR*

The Hard-Wired Brain

Right Clears possess strong traits of logical and progressive thinking, regard for order and truth, verbal expression and attention to detail. In addition to these traits, in a subordinate and complementary role, are the qualities of the *Clear* side. These give *Right Clears* some ability to access both sides of the brain, the verbal and the visual, the linear and the spatial, without letting emotion get in the way of their assessment of a situation.

What this combination of traits does for the *Right Clear* is to give them a touch of extra creativity, even non-conformity in their step-by-step approach to problem-solving. It is an imaginative side of brain behavior, operating in linear and verbal modes. We tend to think of creativity as confined to the visual, abstract and the intuitive, capable of far-reaching innovation. The *Right Clear's* creativity may take longer, because their linear and factual traits are less inclined to leaps of faith and hunches than the creative side. There will be less fanfare and fireworks when a *Right Clear* creates something, even if it is no less earth-shattering than some outrageous *Left Double's* discovery.

With the *Right's* grasp of detail and the *Clear's* adaptability, the *Right Clear* has a competitive edge over the other BrainLines readings, which is very important to both the *Clear* and the *Right* parts of the reading. *Right Clears* have a degree of sensitivity towards others which will keep them from going all out for the prize with no holds barred.

As is the case with all readings with BrainLines, *Right Clears* are aware of constant input through all six of their senses.

Emphasis on logic and clarity makes them try to bypass emotional issues, which, to them, only make life unnecessarily complicated and confusing.

Feeling the Feelings

Right Clears will often be thought of as distant and detached in their dealings with other people. This, however, disregards the depth of their empathy towards others. They simply do not like to use emotional energy in their decision-making and prefer it when others follow suit. In social situations involving the people with whom they are most comfortable they are usually friendly, but guardedly so.

To the typical *Right Clear*, life is a procession of details that must be arranged to prevent possible trouble. The only way for them to get through life is to use their practical and diagnostic skills without letting feelings get the upper hand. They are always on the alert for trouble and trying to be one step ahead of it by planning for every contingency. There is a danger of this making the *Right Clear* lean toward pessimism as a buffer between them and disappointment.

The fact that they are less overtly emotional than some others does not mean they ignore the needs of the people around them. They are more likely to express their empathy by suggesting well-thought-out solutions to problems that may be causing distress to others. At the time, this may seem a cold and clinical attitude to the people they are trying to help, but the fact that they are willing to spend time and energy on the situation is the *Right Clear's* way of showing that they care. They are thus capable of showing their love for their friends and associates, as long as it can be channeled in a practical way.

Above all, they value facts rather than feelings to get through life. Serious non-fiction is typically their chosen reading, especially if it sheds some light on a situation that is currently affecting them. Poetry and abstract art are unlikely to appeal to *Left Clears*, who like to get to a recognizable point as soon as possible with no euphemistic and representational devices.

Social Ability

Because of the confidence they have in their own opinions, solutions and habits, *Right Clears* prefer to either work alone or assume a leadership role. They will often find people with other BrainLines to be exasperatingly illogical. They will have to make a conscious effort to make team-members feel that they are contributing by at least appearing to consider their ideas. Their verbal skills are such that they are articulate and persuasive enough that they can inspire others to follow their lead. The hardest part of team-work for them is the need for showing concern for the feelings of others, an emotional interruption. Their sensitivity and loyalty helps them accomplish this, whether they recognize it consciously or not. Personal rapport is not easy for them, but not impossible.

Their perceptive nature, pragmatism, and conscientiousness make it difficult for them to be anything but honest in their dealings with others. Combined with their desire for accuracy and their practical abilities, this makes them trustworthy and reliable in all their endeavors.

Right Clears have little time for small-talk, preferring clarity of expression and pointed discussion to chattiness. They may be impatient when dealing with people who are not as quick to see the salient details of an argument or will not use reason and logic to make their point.

Hunches and intuition translate as ill-informed guesswork to the *Right Clear*. They will try to inject cold, hard truth, as they see it, into any discussion, which, to their amazement and frustration, is not always what other people want to hear. It takes a great effort of will for the *Right Clear* to play along with any exploration of possibilities or discussion of philosophical theories if there is a lack of serious data to support the arguments. *Right Clear's* have to delve deep into their store of patience and perceptiveness to join in that kind of banter, which probably will only seem like a waste of time to them. They may try to join in to humor their friends, but are more comfortable when they can disseminate factual experience and considered opinions.

Right Clears enjoy challenges and competition and are generally self-sufficient. When there is a decision to be made that is not theirs to make, they will be the first to make the need for that decision known. This makes them a voice for progress in the endeavors in which they choose to involve themselves, especially if they feel that there is something interfering with the logical advancement of a project or the proper dissemination of facts which are pertinent to a situation.

If they are involved in something they can truly believe in, they will stand up to any challenge to that belief. They will also challenge any threat to their preferred way of life. This is due to the *Clear* side's focus and instinct for survival. For all their pragmatism and survival skills, *Right Clears* are still more vulnerable to emotional pain than a true *Clear*.

Right Clears are not suited for situations in which diplomacy and tact are the foremost modes of communication, except, perhaps, in situations where they see a direct risk to their personal survival. The ambiguities involved in trying to avoid confrontation by bending the truth to suit all those involved are repugnant to *Right Clears*, who value their truth above all. The political world would not hold much fascination for them, unless they were involved in a cause about which they felt strongly enough that they could overcome their distaste for spin-doctoring the truth.

Dating and Relating

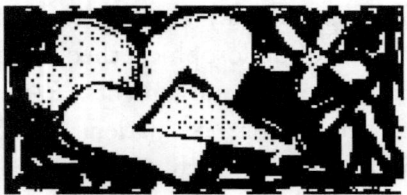

A *Right Clear* person is an objective and faithful partner in a personal relationship, generally preferring to be in a stable, long-term commitment. Their neat and logical way of looking at things makes relationships full of emotional danger for the pragmatic but covertly sensitive *Right Clear*. They are as conscientious in trying to make a relationship work as they are in making any system run smoothly. They will stand by their decisions and goals, and once they have decided to invest time and energy into something, including a relationship, they will do whatever they can to try to make it work. There is an element of stubbornness in this attitude that can be overcome by an appeal to their inner sensitivity.

If their partner is willing to have their relationship planned for them, the *Right Clear* will take charge. This may mean that, once the relationship is on a steady footing and seems to be running smoothly, the *Right Clear's* attention may be diverted to another project or problem to be solved. Relationships and their challenges are important to *Right Clears*, even though they may try to approach them in an unemotional way. Their feelings are deep and they prefer them to be guarded to preserve their privacy as much as possible, but this does not mean they do not feel as strongly about a relationship as their partner.

A relationship with a *Center* would set a tone of emotional communication in which romance, empathy and close companionship could flourish, as well as heightening the *Right Clear's* esthetic appreciation.

Relationships between *Right Clears* and their opposites, the *Left Clears*, benefit from the increased intimacy and emotional assertiveness that the *Left Clears* bring, while contributing their strong ability to guide and to focus on different kinds of subjects. This could create an ideal, creative, emotional collaboration, using innovative logic and intuitive imagination.

Other choices would be either a *Balanced Double, Left Double* or *Clear*. The first two would more than satisfy the *Right Clear's* emotional needs. They would respect and appreciate the *Right Clear's* deductive and imaginative approach to all matters, even if they did not understand it. *Right Clears* would lend their stability to a relationship with a *Balanced Double* or *Left Double*.

Clears would be just as compatible, but without the emotional demands for intimacy that the *Balanced Doubles* and *Left Doubles* would make.

A *Left Clear* or *Left Double* person would probably have to initiate the first encounter and eventual intimacy with a *Right Clear*, because the *Right Clear* depends on previous experience and has difficulty in starting new personal relationships. The *Right Clears'* tendency towards the straightforward approach can be tempered by the more romantically-inclined right-brained partner.

Intimacy would probably take a back seat in a relationship with another *Right Clear*. Although they would be very compatible, they would have to agree not to expect to spend very much private time together. Work would tend to be their emphasis.

Despite their verbal skills and ability to make quick decisions, when confronted by their own romantic impulses *Right Clears* may feel that they are in over their heads. Their normal style is to state facts logically, but when trying to express love in a logical way, the meaning tends to get lost in the translation. While they may be able to express their understanding of the state of love, they may find it hard to share their feelings about it. Partners may find themselves asking that their *Right Clear* companion tell them at least once a week that they love them, which is one way to get the

Right Clear to declare their feelings and, at the same time, feel a sense of order.

The *Right Clears'* way of showing affection is often in the things they do, rather than what they say. They need to be reassured that this is a valid expression of their love. As an indicator of the depth of their devotion, they will try to make home life run as smoothly and efficiently as they can, in order to keep the load off their spouse's shoulders. Pleasure and work will be organized and scheduled, punctuality will be the norm and honesty the byword. They will work hard at keeping their family happy, by doing little things round the house and feeling rewarded by the gratitude they hear in their spouse's voice. For some partners, this will be enough, others may want more emotional shows of affection, but these may only be achieved by pleading for them, with the danger that any resulting words of love may be more by rote than expressed with feeling.

Relationships with *Right Clears* tend to grow through discussion of problems or common dreams, rather than by trusting that things are and will be all right. In every subject but the expression of feelings, *Right Clears* are verbal thinkers. They will want to talk everything out, which is the way to get problems on the table. They have no fear of words and, once they are on a roll, they will make decisions based on the information they receive, which means that their partner must be as open as possible in any talks they have.

Compatibilities

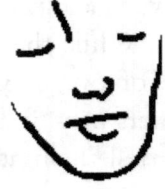 What are you looking for in a romantic relationship? Here are the major qualities to expect in a partnership when a *Right Clear* pairs up with one of the other BrainLines readings.

Right Clear with Center
Intellectual stretching, if *Right Clear* tries to approach *Center's* holistic view.
Stability in well-defined relationship a priority.
Room for each other's pursuits.
Conscious effort required on both sides to initiate intimacy.

Right Clear with Balanced Double, Left Double or Right Double
Emotionally charged relationship may be challenging to *Right Clear*.
Romance is a priority for *Balanced Doubles*, *Left Doubles*, and *Right Doubles*, which may be difficult for the *Right Clear* to sustain. Spontaneity would give an extra dimension to the *Right Clear's* life.

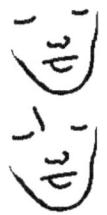

Right Clear with Clear or Right Clear.
Low emotional level most comfortable for *Right Clear*.
Empathy for each other's styles.
Career focus is understood by all these readings.
Commitment to a stable relationship is a priority.

Right Clear with Left Clear.
Creativity from both partners will increase with mutual encouragement.
Intimacy increases as different styles mesh.

Similar focus on different interests.
Rounded and stable relationship.

Work That Works

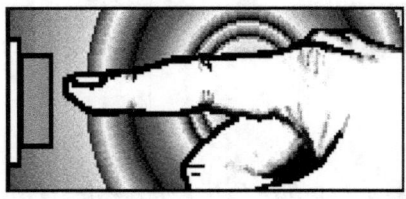 *Right Clears* are most comfortable in jobs that make use of their verbal skills, their strengths in logical analysis, linear creativity and their quick decision-making. They are responsible and conscientious in their work and show great patience with the details of a project. In setting up and completing projects they are focused and organized. They are well-suited to systems-design work. *Right Clears* will work through even tedious jobs in an unruffled and meticulous way until they are done to their satisfaction.

Right Clears form opinions and make choices that are unaffected by emotional constraints. They state the facts as they see them, give their opinions and are loth to allow their judgment to be affected by what others may think or feel.

They are good at making quick decisions based on factual information and with tangible goals. Once set on a course of action with a set goal, they are willing to change tactics midstream if it seems appropriate, but unwilling to change course completely. If the goals of a project are changed by others, *Right Clears* will have a hard time adapting to a whole new set of criteria.

They will not pursue a goal with no thought of how others will be affected. They simply believe that emotions have no place in the logical analysis of existing data. *Right Clears* share the sensitivity of all the readings with BrainLines to the emotional health of the people around them, but feel that there is a time and place for everything. Emotions by themselves must not be allowed to override practical considerations, but when *Right Clears* are cognizant of actions that may have harmful repercussions, they will typically try to avoid them.

Right Clears prefer to work alone, because they are uncomfortable around people whose working modes are different from theirs. They can be mystified by what they perceive as the irrational actions of those who are more right-brained. Intuition, to *Right Clears*, is just another word for pure guesswork, with no basis in reality. They tend to overload themselves with work, because they distrust the methods of others and so are hesitant to delegate responsibility. If a *Right Clear* is asked to join a team, it should be made plain that he or she will have to recognize the strengths of the other members. If *Right Clears* can be taught to trust other people's skills, they may be able to bring their logic and decisiveness to the work without taking too much upon themselves.

Generally, *Right Clears* are not made to be people-managers. They are more suited to planning courses of action, preventing crises with a cool head and designing logical systems that can handle any foreseeable eventuality. The 'foreseeable' can, however, be distorted by the whims of other people who come into contact with the system. That is the kind of eventuality that will confound *Right Clears*. But, as the original system crashes about their ears, they will be examining the new parameters and formulating a new version to deal with the new reality.

When presenting their ideas, *Right Clears* will announce them as undeniable facts. Once they have formed opinions or come up with plans, they will make a stand against any challenge, until seeing the value of the argument. If they are in charge, this may lead to a working environment in which other people feel browbeaten by the *Right Clears*, who, at first, will see other points of view as simply the inability of other people to keep up with their rapid, logical thinking. They will argue their point, may appear combative, but they are really just trying to make others see their reasoning.

In confrontational situations *Right Clears* will be the voice of reason, backing up every statement with solid fact, standing steady against any emotional barrage that may be directed at them. This quality can be very useful in fields that require face-to-face discussion in which facts, stated calmly and irrefutably, will sell the product, win the case or get a signature on the contract or treaty.

Since they hesitate to share decision-making, they are likely to be in fields in which they can operate with the minimum of social interplay. They can, however, turn on the personal charm if a project, their business or their own survival is at stake. They have the ability to provide a confident analysis of facts, while focusing on what needs to be done to win a client over, without distorting the truth or giving up too much.

Right Clears seek precision and creative innovation in everything, so any field in which that is important would appeal to them. Accounting, photography and other technical arts such as glasswork and jewelry, mechanical and computer systems, software development, diagnostic medicine, laboratory research, and corporate law are all natural environments for *Right Clears*, along with any field in which mathematics or assessment of data form most of the work load.

Right Clears like to keep matters simple and straightforward, sometimes preferring to limit the input on a subject rather than complicate it with possibly conflicting information. They like to have 'either/or' choices, not multiples, so that their rapid judgment skills can come into play. Once a project is underway, *Right Clears* prefer to deal with the fundamentals, troubleshooting the system rather than the people involved in it. They will want to concentrate on completion of the project, bringing their intense focus to bear on it and hoping that everyone does his or her part to work towards that goal.

Right Clears are good at keeping meetings on track, forming and sticking to the agenda and stating the pitfalls that they can see in a plan. While they may lack some interpersonal skills and may be stubborn about the plans they devise, they will be ready with new ideas as soon as the new criteria are made clear to them. These are people who are turned on by the challenge of making things work efficiently and quickly. It is important that a mutual understanding is attained between *Right Clears* and their workmates of their respective strengths and what they can bring to a project.

BrainLines of the Rich and Famous
Right Clear

Orlando Bloom, James Broughton, Robert Carlyle, John Cassavetes, Miguel Ferrer, Madeline Gleason, William Hurt, Lee Iacocca, Rick Kaplan, Ralph Lauren, Michael McClure, Yevgeni Primakov, Queen Elizabeth II, Adlai Stevenson, Eric Stoltz.

PERSONALITY PROFILE FOR *RIGHT DOUBLE*

The Right Double *displays two lines, one on each side of the forehead between the eyebrows. The line on the right-hand side is longer than the one on the left-hand side.*

Quick Summary for *Right Double*

Read this to get a feel for what BrainLines can tell you about yourself and people you meet. The Full Profile begins on the following page.

You're the observer with the twinkling eye and the concise one-liner to sum up the occasion. You could be a rocket scientist with an art degree. Although you would love to organize your life, you have that quirky side that keeps everything slightly off-balance. Don't deny your sensitivity to the feelings of the people around you. It's a valuable asset. Thank goodness you're around to give sensible advice to the more illogical among us, yet maintain your sense of humor. If you were an IPO, you'd be a great investment.

Relationship tip: You are tolerant, idealistic, rational and practical. Need we say more? Well, O.K., you're also prone to letting your career take precedence over your relationship. But you have some built-in emotional radar that will tell you when your partner's getting antsy for expressions of love. Pay attention and don't keep your love-life on auto-pilot.

FULL PROFILE FOR *RIGHT DOUBLE*

The Hard-Wired Brain

 *Right Doubles* approach the dichotomy of their left and right influences in a unique way. They accommodate the creative, idealistic and visionary side of themselves in a progressive, organizational and methodical way. This means that their linear-thinking dominant side is affected by the quirky originality of their abstract/emotional side, giving them unique methods of dealing with life.

Their uniqueness makes itself felt through their own brand of creativity. We tend to think of creativity as confined to abstract/intuitive artistic talent. Creativity is just as capable of far-reaching innovation, even if it operates in linear and verbal modes. Because linear creativity cannot work on hunches, it takes longer but has no less potential to change the world.

Right Doubles are always observing and analyzing the interactions around them. They are modest and detached onlookers, friendly but somehow distant. While they may not seem to be taking much of an interest in what is going on around them, they can suddenly come out with a concise one-liner that will sum up the entire proceedings.

That combination of precision and cool amusement really sums up the *Right Doubles*. They are people to whom the business of making choices and forming opinions is best based on the available facts. They can make quick decisions that are concise, practical and to the point. In their immediate responses to situations, they plan for everything to run smoothly along well-ordered and pre-determined lines, with no disturbance in the smooth transition from the present to the future, using past experience as a guide. Then, rushing in from their blind side, comes their abstract influence, a little out of breath and apologetic for being late, with spontaneous emotional responses, sparkling images and suggestions. They can thus temper their somewhat stodgy serious-mindedness and can laugh at themselves while carrying out the plans they make.

Right Doubles are such original thinkers in the ways they find for improving techniques or finding innovative ways of using existing systems, that they sometimes find themselves out of step with the prevailing norms in their field. This is because of their quirky imbalance which may make their ideas too original to be accepted in the present.

They will be able to focus on short-term repetitive tasks if the goal is important enough and satisfies a logical need in the creation of something important to them. In this kind of situation they will work calmly and steadily as long as the end is in sight. *Right Doubles* have the tolerance and flexibility to cope with adverse situations, bearing them stoically and absorbing themselves in their own concise and thoughtful actions. They do not tolerate long-term repetitive situations well, but will, if possible, delegate them to others more adaptable to such tasks.

They are interested in analyzing the meaning of life and other matters concerning the spirit and the soul, perhaps more as an exercise in investigation than as a true yearning for inner understanding. *Right Doubles* are likely to be uncomfortable with organized religion which may require illogical leaps of faith, although they may be initially drawn by the guidelines and rules. While they are tolerant and flexible enough to understand other people's need for direction in their spiritual search, *Right Doubles* are such original thinkers that they will prefer their own methods. They form their own carefully considered theories about the meaning of life and are comfortable with their own form of spirituality.

Feeling the Feelings

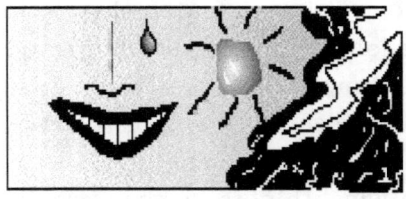

Right Doubles have an insight into others' feelings that they share with the opposing *Left Double.* They are emotional people, but do not readily lay their souls bare. The detachment that masks their hidden depths is their buffer against the vulnerability created by their sensitivity. This distancing of themselves from outward display of emotion helps them to stay focused on their actions with an air of quiet concentration and self-confidence. They may not be feeling that way on the inside, however.

This emotional and sympathetic range means that they have many thoughts vying for attention, some of which may involve them in social or community problems and solutions. *Right Doubles* will deal with these in an organized way, getting one part accomplished before moving on to the next part, but the awareness they have of other people's feelings, and the urge to help them, may distract them for a while. Once again, their creative influence has disturbed their orderly way of getting on with business. Feeling slightly irritated, they will attend sincerely to the task of helping a friend in need. Then they will draw themselves back in and concentrate on the more rational task at hand, basking in the joy of using their logical talents.

Social Ability

Right Doubles can use their sensitivity to the feelings of those around them to create structured responses to problems. They can often be valued advisors to their friends, who will benefit from their clear-headed, sympathetic viewpoints and solutions. This kind of reasoning, which comes from both sides of the brain, can take the *Right Double* into fields where quick decisions along logical lines are of value, where those decisions are tempered by regard to their effects on the people involved, or where a little intuition is sometimes necessary. Anything that requires stifling their social conscience will be damaging to *Right Doubles*.

In common with their fellow *Doubles,* they show the tendency to have many interests, preferring occupations and hobbies with established methods or sets of rules within which they can exercise their creativity. For this reason, *Right Doubles* tend to gravitate towards the more technique-oriented creative pursuits.

Right Doubles have their intuitive side which is helpful to them both in the formulation of their religious or spiritual beliefs and in their personal interactions. Their sensitivity to others' emotions may make them uncomfortable at times, but they cannot help but be attuned to other people's welfare.

In politics, which is a field where they could be a reasoned voice for forward thinking and rational policy, they would tend toward altruism. The belief in order would keep them honest through respect of the system, although they would not be averse to modifying the system where it was flawed.

Right Doubles do well in careers requiring snap decisions with people's futures in mind. They are adaptable enough that they are able to consider all sides of a question and come up with a carefully considered answer that attempts to address everyone's concerns. Patience is also one of their virtues, which can be useful when dealing with people with radically different points of view.

Their organizational skills coupled with their adventurous spark make *Right Doubles* perfect for any kind of proactive work. Being involved in issues that affect their neighbors' lives would be challenging and fulfilling. Dealing with social and political questions requires the ability to think on one's feet, in the full realization of the consequences. That would be a delight for the self-actualized *Right Double* and would use the imaginative skills of both brain hemispheres.

Dating and Relating

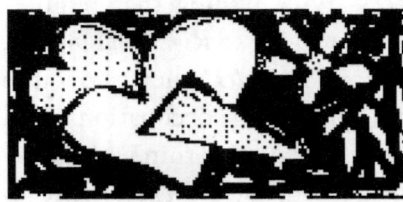

Right Doubles develop relationships fairly slowly. They are as careful in choosing the people with whom they become intimate as they are in selecting the words they use to express their feelings. They are not shy about their emotions, but they do not like to express them without due consideration of the truth of what they feel.

Their word is their bond (no less in professing love for someone than in a business deal), to the extent that they may try to save a relationship that has soured beyond the point of no return. They are often saved from such situations by yearning for constant improvement in a partnership and the need for adequate return of their affection. If they listen to their proddings, they will realize when a relationship is failing beyond hope and perhaps save themselves some emotional scars.

For the typical *Right Double*, setting aside the time from their career to begin a possible romance is a big step in itself, which should not be taken lightly by their prospective partner. Their tendency is to put work first and play second. Commitment to their chosen career will always be a pull for them. Their sensitivity to their partner's emotions will usually tell them when feelings of neglect are arising, but *Right Doubles* may sometimes disregard their emotional 'radar', especially in times of work-related stress.

Right Doubles are natural organizers and this is true even in their romantic relationships. When there is time for leisure activities, they will be scheduled for maximum efficiency. Partners of *Right Doubles* should expect to have their free time planned for them and to sometimes feel that they are operating under scheduled categories of fun, work, romance, travel time, etc. This makes it all the more delightful when, during one of the *Right Double's* planned activities, they will upset the schedule by suggesting something completely unexpected and goofy. It is that quality of serendipity lurking beneath the ordered exterior of *Right Doubles* that makes them interesting and amusing.

The partner who might enrich a *Right Double's* life most and would appreciate their wacky side would be a *Left Double*. These partners would be a continuing source of fascination and entertainment, who would contribute to and encourage the *Right Double's* creative and emotional propensities. Such a partnership would require effort on the *Right Double's* part to stabilize it and pursue a long term relationship. These two readings share the ability to access both sides of the brain, but start from opposite sides. Their initial reactions to everything would therefore be experienced differently. Given time and perseverance, however, they could find their common ground and learn to appreciate their respective strengths.

If a *Right Double* were pursuing a less emotional, more logical direction in a relationship, they would probably prefer a *Right Clear* or a *Clear*, who would perhaps collaborate and would definitely relate to their methodical and linear-thinking side. A long term relationship with either of these readings would be easier for them than with anyone with the highly emotional readings. Such a relationship would perhaps not exercise the *Right Clear's* traits to their fullest, creating a domestic atmosphere that might be somewhat cool and that perhaps became secondary to their workplace.

A relationship with a *Center* would provide the stimulation of being with someone who shared the *Right Double's* propensities for access to both sides, yet would give them the room to pursue their own interests. The *Centers* would be able to bolster the *Right Doubles'* strengths, encourage their weaker traits, but would not require constant positive reinforcement, thanks to the *Centers'* emotional self-sufficiency. *Right Doubles* in such a relationship would find themselves in the unexpected position of being more romantically demonstrative than their partners.

Right Doubles may be the quiet ones who like to stick to their own circle of tried and true friends, but once you have their attention and they are attracted, they may surprise you with their occasional flights of fancy. Most of the time, they will try to keep everything on as even a keel as possible, then, in order to show how much they care, they will do something unexpected. They are just as likely to offer to fix your car as they are to leave a poem on your answering machine. Both of these convey their love, and they sum up the *Right Double's* practicality and quirkiness.

Right Doubles make good partners in marriage because of their ability to be good-humored, tolerant, idealistic and practical. They are open to any logical method of dealing with problems, emotional or otherwise, and will work creatively for a successful relationship. Partners and family members will find them to be rational and caring, always ready to give advice quickly, in a calm and considered voice of experience.

Compatibilities

What are you looking for in a romantic relationship? Here are the major qualities to expect in a partnership when a **Right Double** pairs up with one of the other BrainLines readings.

Right Double with Center
Intellectual stimulation.
Creative inspiration.
Reinforcement of *Right Double's* strengths.

Time for each other's own interests.

Right Double with Balanced Double, Left Clear or Left Double
High emotional level could be a challenge.
Unpredictability adds spice to the relationship.
Different styles require compromise from both partners.
Complementary traits aid creativity.

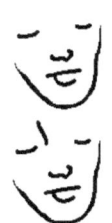

Right Double with Clear or Right Clear
Predictability and stability of relationship.
Focus on careers a priority.
Low emotional levels may mean working on expressions of love.
Right Double will need space for his/her own interests.

Right Double with Right Double
Deep empathy.
Collaboration on all levels.
Easy relationship, with defined roles and room for quirkiness.

Work That Works

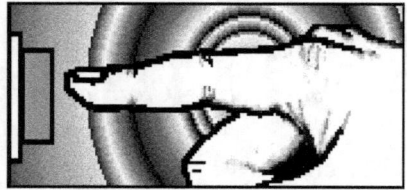 *Right Doubles* would like everything to work in a structured, logical and timely fashion and they will work towards that goal. Their brain functions are such, however, that very soon after they have devised a logical system with which to work, they may come up with emotional objections and slightly off-the-wall ideas.

Right Doubles are very aware of the emotional states of the people around them, which can be a distraction from their ordered way of working. Their concern for others may cause them to modify systems they have devised in order to keep everyone as happy as possible.

They are flexible in their attitudes, although they prefer that their life be planned as far ahead as possible. Once they are given goals to reach, they will devise long-term scenarios, trying to think of all the pitfalls and planning for every eventuality. Their strength is the ability to think along more intuitive lines, especially where planning for people's reactions is concerned. This gives them an edge in dealing with customer satisfaction with new products and services. If they know the people with whom they must work, *Right Doubles* will be able to get a feel for their personalities and form strategies to deal with them.

They are especially good at devising teamwork situations, once the goals and guidelines have been set up. They will be able to establish a rapport with different kinds of people, deal with their work styles, yet maintain a logical progression in their chosen course of action. *Right Doubles* will try to see all sides of a question, should objections to their procedures arise, which helps them to visualize new ways of dealing with problems and keeps them from being stubborn about their ideas. They are able to see the importance of the strengths that other team members possess and will be fairly comfortable in delegating tasks and even decisions to them.

Their analytical strengths give them confidence in their own decision-making. They explore the downside of any course of action, and are prepared to take a moderate risk, if they feel it is justified in reaching the goals they have set. With their attention to detail and their penchant for organization, they minimize the risks they have to take, but they realize that life is full of unknowns and that, no matter how they may wish to, they can never foresee every problem that will arise. They are not too surprised when things start going in a different direction from what they had planned and, with any luck, will have had the foresight to give themselves maneuvering room. If they feel good about the project, they will be willing to do whatever it takes to deal with obstacles, using their dual sides of creativity.

Areas most suitable for *Right Doubles* are those which allow them to use their logical and organizational abilities, their interpersonal skills and their creativity. Management at all levels will use their strengths. Their team-building skills and their preference for regulation and order make them suitable for government or military careers. *Right Doubles* would make good strategists in the field, with their logic and flexibility to fuel innovative concepts. *Right Doubles* could also use their talents in the technology field, where plans for product development and marketing must be made quickly in order to keep up with the volatile nature of the industry.

If *Right Doubles* enter the art world, they are most likely to be found in disciplines that appeal to their technical abilities. Fields such as photography or jewelry-making, in which the technique creates rigid parameters within which they can express themselves would probably appeal more than abstract forms. Their aptitude for language would also make them well-suited for writing of all kinds, as would the formalized structure of musical composition and musicianship. That sort of mixture of the technical and the imaginative appeals to the *Right Double* as an outlet for their innate abilities.

Where their decisions are concerned, *Right Doubles* will have an emotional attachment and will tend to bristle a little at criticism. They, too, may see the logic in another answer to a problem, but that touch of emotion means they need positive feedback on a job well done.

The unevenness of their propensities can make them feel out of balance, especially if they are dealing with an unfamiliar situation. *Right Doubles* who have not come to terms with the sometimes troublesome creative input, which usually kicks in just when the linear has everything figured out, can find themselves suppressing it or ignoring it. This may create a lack of self-confidence that can be detrimental to their performance. They need to be reassured that both traits are equally important to their life and work.

Self-assured *Right Doubles* are valuable for their logical analysis of facts and situations, their ability to condense data and make quick decisions and for their sensitivity and willingness to bring other people's skills to a project. They respect honesty and straight talk, preferring not to beat around the bush themselves, but will be prepared to wait out the digressions of other people if they think they have something to add to the discussion. They will inspire others with their enthusiasm for the work at hand and command respect for their understanding of the ways others like to work.

BrainLines of the Rich and Famous

Right Double:

James Baldwin, Mayor Willie Brown, Jim Carrey, Julia Childs, Chow Yun Fat, Peter Coyote, Leonardo DiCaprio, James Dickey, Michael Douglas, Albert Einstein, Charlton Heston, Don Johnson,

Jack Kevorkian, Kevin Kline, D.J. Mendel, Gabriel Orozco, Leon Panetta, Robert Redford, Kenneth Rexroth, Geraldo Rivera, Mister Rogers, Orville Schell, Charles Schwab, Gen. Norman Schwarzkopf, Sarah Silverman, Jessica Tandy, Elizabeth Taylor, Ted Turner.

BIBLIOGRAPHY

Abrams, Jeremiah (ed.), Thomas Moore (ed.); *The Shadow in America: Reclaiming the Soul of a Nation*: (1994); Nataraj Publishing.

Armstrong, Thomas; *7 Kinds of Smart: Identifying and Developing Your Many Intelligences*: (1993); Plume.

Aron, Elaine N.; *The Highly Sensitive Person: How to Thrive When the World Overwhelms You*: (1996); Birch Lane Press.

Asano, Hachiro; *Faces Never Lie*: (1964); Rikugei Publishing House.

Berland, K.J.H; *Reading Character In The Face; Lavater, Socrates and Physiognomy*: (1993); Word And Image, v9, no. 3, July-Sept. 1993.

Cohen, Ronald Jay; Montague, Pamela; Nathanson, Linda Sue; Swerdlik, Mark E; *Psychological Testing: An Introduction to Tests and Measurement*: (1988); Mayfield Publishing Company.

Cohen, Ronald Jay; *101 Exercises in Psychological Testing and Assessment, Third Edition*: (1996); Mayfield Publishing Company.

Cutter, Rebecca; *When Opposites Attract: Right Brain/Left Brain Relationships and How to Make Them Work*: (1994); Dutton.

Descartes, René; *Passions Of The Soul*: (1649). Translated in *Body and Mind: Readings in Philosophy*: (1964); G.N.A. Vesey, ed.; George Allen and Unwin.

Fitzherbert, Andrew; *The Palmist's Companion: A History and Bibliography of Palmistry*: (1992); Scarecrow Press.

Fosbrooke, Gerald Elton; *Character Reading Through Analysis of the Features*: (1914); G.P. Putnam's Sons/Knickerbocker Press.

Frager, Robert, ed. *Who Am I? Personality Types for Self-Discovery*: (1994); Jeremy P. Tarcher/Putnam.

Friedlander, Joel; *Body Types: The Enneagram of Essence Types*. (1986); Inner Journey Books.

Furnham, Adrian and Stringfield, Paul; *Personality and Occupational Behavior: Myers-Briggs Type Indicator Correlates of Managerial Practices in Two Cultures*: (1993); Human Relations, vol. 46, no. 7, p.827.

Garrison, Omar V; *Tantra: The Yoga of Sex*: (1964); The Julian Press.

Gazzaniga, Michael S.; *Mind Matters: How Mind and Brain Interact to Create Our Conscious Lives*: (1988); Houghton Mifflin Company.

Goldberg, Philip; *The Intuitive Edge: Understanding and Developing Intuition*: (1983); Jeremy P. Tarcher, Inc.

Goleman, Daniel; *Emotional Intelligence: Why It Can Matter More Than IQ*: (1995); Bantam Books.

Hall, Calvin S. and Nordby, Vernon J; *A Primer of Jungian Psychology*: (1973); Mentor.

Hamer, Dean and Copeland, Peter; *Living With Our Genes: Why They Matter More Than You Think*: (1998); Doubleday.

Harrington, Anne; *Medicine, Mind, and the Double Brain*: (1987); Princeton University Press.

Hipskind, Judith; *Palmistry: The Whole View:* (1988); Llewellyn Publications.

Hipskind, Judith; *The New Palmistry:* (1994); Llewellyn Publications.

Jung, C.G; *Collected Works, Vol. 6*; Psychological Types: (1953); Princeton University Press.

Kurtz, Ron and Prestera, Hector; *The Body Reveals*: (1976); Harper and Row/Quicksilver Books

Kushi, Michio; Oriental Diagnosis: *What Your Face Reveals*: (1978); Sunwheel Publications.

Leadbeater, C.W; *The Chakras*: (1971); The Theosophical Publishing House.

Mar, Timothy T; *Face Reading: The Chinese Art of Physiognomy*: (1974); Dodd, Mead and Co.

Marcia, D., Aiuppa, T. and Watson, J; *Personality Type, Organizational Norms and Self-esteem*: (1989); Psychological Reports, No.65, p. 915.

Maslow, Abraham H.; *Toward a Psychology of Being:* (1962); D. Van Nostrand Company, Inc.

Miller, Marlane; *BrainStyles: Change Your Life Without Changing Who You Are*: (1997); Simon & Schuster.

Mishra, Rammurti S; *Fundamentals of Yoga*: (1959); The Julian Press.

Mitchell, M.E.; *How To Read The Language of the Face*: (1968); Macmillan.

McGilchrist, Iain; *The Master and his Emissary: The Divided Brain and the Making of the Western World:* (2009); Yale University Press.

Myers, Isabel Briggs; *Gifts Differing*: (1980); Consulting Psychologists Press.

Myss, Caroline; *Anatomy of the Spirit:* (1996); Three Rivers Press.

Neubauer, Peter and Neubauer, Alexander; *Nature's Thumbprint: The New Genetics of Personality*: (1990); Addison-Wesley Publishing Company.

Ornstein, Robert; *Multimind: A New Way of Looking At Human Behavior*: (1986); Houghton Mifflin.

Ornstein, Robert; *The Roots of the Self*: (1995); Harper San Francisco.

Pelton, Robert W; *Lost Secrets of Astrology*: (1973); Nash Publishing.

Puccetti, Ronald; *Mind With A Double Brain*: (1993) British Journal for the Philosophy of Science, v44, no.4 (Dec. 1993); Oxford University Press.

Relethford, J.H; *Craniometric Variation Among Modern Human Populations*: (1994); American Journal of Physical Anthropology, Sept. 1994; 95(1): 53-62.

Rifkin, Jeremy; *Patent Pending*: (1998); Mother Jones, June 1998; Vol. 23, Issue 3.

Rowe, David C; *The Limits of Family Influence*: (1994); Guilford Press.

Rutter, Michael and Marjorie; *Developing Minds*: (1993); Basic Books

Shiflett, Samuel C; *Validity Evidence for the Myers-Briggs Type Indicator as a Measure of Hemispheric Dominance*: (1989); Educational and Psychological Measurement, vol.49, no.3, p.741.

Siever, Larry J. and Frucht, William; *The New View of Self: How Genes and Neurotransmitters Shape Your Mind, Your Personality and Your Mental Health*: (1997); MacMillan General Reference.

Shlain, Leonard; *The Alphabet Versus the Goddess:* (1998); Penguin Arkana.

Spoto, Angelo; *Jung's Typology In Perspective*: (1989); Sigo Press.

Springer, Sally P. and Deutsch, Georg; *Left Brain, Right Brain*: (1981); W.H. Freeman and Co.

Taggart, William M., Kroeck, K. Galen and Escoffier, Marcel R; *Validity Evidence for the Myers-Briggs Type Indicator as a Measure of Hemispheric Dominance: Another View*: (1991); Educational and Psychological Measurement, vol. 51, no. 3, p.775.

Thakkur, Chandrasekhar; *Your Palm – Your Mirror*: (1980): ©Author.

Tickle, Naomi R.; *It's All In The Face: The Facts and Fantasies of Face Reading*: (1995); Daniels Publishing.

Whiteside, Robert L.; *Personology, The Dynamics of Success*: (1969): Frederick Fell, Inc.

Wigan, Arthur Ladbroke; *The Duality Of The Mind*: (1844/1985); Joseph Simon

Williams. Linda Verlee; *Teaching For The Two-sided Mind*: (1983); Prentice-Hall.

ABOUT THE AUTHORS

Donna Sheehan and Paul Reffell have collaborated on many projects in their twenty-year creative partnership. They share a deep fascination with the human condition, with "Cultural Potholes" – the unseen or ignored parts of human existence that cause so much damage – and with the differences that have evolved between men and women. They are proud to call themselves Evolutionary Behaviorists.

BrainLines was one of their first collaborations and the foundation of their later work together. Their investigation of these physical markers of brain hemisphere influence, building on the decades of Donna's previous observations, continued through the years, as they used their knowledge in all their interactions with other people. They found BrainLines to be so useful that they knew they had to share their findings in this book.

Donna & Paul's other work includes environmental activism (www.MowAndSow.org); peace activism involving women (www.BaringWitness.org); using Darwinian sexual selection theory to teach women about their innate ability to select and guide their mates (www.SeductionRedefined.com), which led to the idea of producing a documentary film (www.TheEveOption.org); and a series of panels, workshops and newspaper articles that bring to light some of the most insidious and damaging Cultural Potholes (www.CulturalPotholes.com).

Donna co-founded a community radio station (www.KWMR.org) and they have co-created interactive art installations – *Wargasm* and *Pro-Degradation*, and an award that honors the great women that stand beside great men – the Dolley Madison Partnership Award. They have been interviewed and featured internationally on television, radio, the Internet and in documentary films.

Please tell friends about BrainLines and send them to
www.BrainLines.com

Please "Like" and share our Facebook page
www.facebook.com/BrainLines

And if you have any insights to share or questions to ask,
please e-mail us at
info@brainlines.com

www.ingramcontent.com/pod-product-compliance
Lightning Source LLC
Chambersburg PA
CBHW051510170526
45166CB00001B/467